Practical Product Assurance Management

Also available for H0966- Practical Product Assurance Management

Also available from ASQ Quality Press

Reliability, Maintainability, and Availability Assessment, Second Edition
Mitchell O. Locks

Failure Mode and Effect Analysis: FMEA From Theory to Execution
D. H. Stamatis

Weibull Analysis
Bryan Dodson

Glossary and Tables for Statistical Quality Control, Third Edition
ASQ Statistics Division

Statistical Process Control Methods for Long and Short Runs, Second Edition
Gary K. Griffith

SPC Essentials and Productivity Improvement: A Manufacturing Approach
William A. Levinson and Frank Tumbelty

Managing the Metrology System, Second Edition
C. Robert Pennella

Practical Product Assurance Management

John Bieda, PE, CRE, CQE

ASQ Quality Press
Milwaukee, Wisconsin

Practical Product Assurance Management
John Bieda, PE, CRE, CQE

Library of Congress Cataloging-in-Publication Data

Bieda, John, 1961-
 Practical product assurance management / by John Bieda.
  p.   cm.
 Includes bibliographical references and index.
 ISBN 0–87389–375–1 (hardcover)
 1. Quality assurance—Management.   I. Title.
TS156.6.B53   1997
658.5'62—dc21   97–27776
                                                 CIP

10 9 8 7 6 5 4 3 2 1

ISBN 0-87389-375-1

Acquisitions Editor: Roger Holloway
Project Editor: Jeanne Bohn

ASQ Mission: To facilitate continuous improvement and increase customer satisfaction by identifying, communicating, and promoting the use of quality principles, concepts, and technologies; and thereby be recognized throughout the world as the leading authority on, and champion for, quality.

Attention: Schools and Corporations
ASQ Quality Press books, audiotapes, videotapes, and software are available at quantity discounts with bulk purchases for business, educational, or instructional use. For information, please contact ASQ Quality Press at 800-248-1946, or write to ASQ Quality Press, P.O. Box 3005, Milwaukee, WI 53201-3005.

For a free copy of the ASQ Quality Press Publications Catalog, including ASQ membership information, call 800-248-1946.

Printed in the United States of America

 Printed on acid-free paper

American Society for Quality

Quality Press
611 East Wisconsin Avenue
Milwaukee, Wisconsin 53202

This book is dedicated to my wife Susan, my son Johnny, my daughter Rachael, and my sister Janine.

Contents

Preface   xi

Introduction   xiii

List of Tables   xiv

List of Figures   xv

Chapter 1   Structure of Product Assurance and Its Management   1

    A.   Definition of Product Assurance   2

    B.   Reasons for Product Assurance   5

    C.   When Product Assurance Is Implemented   6

    D.   Where Product Assurance Is Applied   8

    E.   Organizational Structure of Product Assurance   9

    F.   Management and Implementation of Product Assurance   9

Chapter 2   Past and Present Perspectives on Product Assurance   13

    A.   Mass Production versus Lean Production Philosophy   14

    B.   Deming's Fourteen Points and Their Relationship to Product Assurance   18

    C.   QS-9000 Requirements and Their Relationship to Product Assurance   21

Chapter 3   Product Assurance Planning within the Product Development Process   23

    A.   Product Development Process (PDP)   24

    B.   Product Assurance Planning (PAP)   24

    C.   Integration of Product Assurance Tools and PAP into the PDP   26

Chapter 4   Product Assurance Tools and Processes and Their Application: Concept Development Phase Tools   32

    A.   Requirement Definition and the Use of Quality Function Deployment (QFD)   33

    B.   Benchmarking   38

    C.   Test Plan Development, Sample Planning, and Analysis   44

        1.   Hardware Model   44

        2.   Software Model   55

Chapter 5   Product Assurance Tools and Processes and Their Application: Design/Process Development Phase Tools   63

    A.   Design and Process Block Diagrams   64

B. Risk Analysis   66
  1. Failure Mode and Effect Analysis (FMEA)   66
  2. Fault Tree Analysis (FTA)   72
C. Design for Manufacturability and Assembly (DFMA)   76
D. Variation Simulation Analysis (VSA)   80
E. Process Control Plan Development   85
F. Measurement System Evaluation   90
  Calibration   91
  Gauge Repeatability and Reproducibility (GRR)—Variable Data Measurement Systems   92
  Gauge Repeatability and Reproducibility (GRR)—Attribute Data Measurement Systems   96
G. Preventive Maintenance   98
  Tool and Equipment Life Studies   100
  Scheduled Preventive Maintenance Plan   100
H. Reliability Growth Management   103
I. Design Review   110

Chapter 6   **Product Assurance Tools and Processes and Their Application: Design/Process Validation Phase Tools   113**

A. Test Data Analysis   114
  1. Variable Type Data   114
  2. Attribute Type Data   127
B. Reliability Test Data Evaluation   131
  1. Stress/Strength Interference   131
  2. Reliability Demonstration Test Analysis   134
  3. Weibull Distribution Analysis Technique   138
  4. System Reliability Block Diagram Analysis   142
C. Quality Engineering Evaluation   148
  1. Process Performance Analysis   148
    Process Performance and Capability Indices: Variable Data   151
    Process Capability Indices: Variable Data   156
    Process Performance Index: Attribute Data   156
  2. Design of Experiments (DOE)   158
D. Process Review   163

Chapter 7   **Product Assurance Tools and Processes and Their Application: Production and Continuous Improvement Phase Tools   173**

A. Statistical Process Control (SPC)   174
B. Pareto Chart Analysis   182
C. Root Cause Analysis (RCA)   184

Chapter 8   **Product Assurance Training   189**

Chapter 9   **Cost of Quality Management   195**
   A.   Cost of Quality Analysis   196
   B.   Life-Cycle Cost Analysis   198
   C.   Product Assurance and the Cost of Quality Management   202

Chapter 10   **Product Assurance Managerial Practices and Behavior   205**

Chapter 11   **Future Direction of Product Assurance   210**
   A.   Growth in Design and Manufacturing   211
   B.   Product Assurance Management in the Service Industry   217
   C.   Growth in the Service Industry   218

Chapter 12   **A Summary of Practical Product Assurance Management   221**
   A.   Benefits   222
   B.   Concluding Remarks   222

Glossary   225

Index   239

Preface

The primary purpose of this book is to educate the reader on how to implement and manage a practical product assurance program in a design and/or manufacturing environment. This book exposes the reader to primary elements of a product assurance process and how to manage the activity successfully. Organizing and explaining the material from a practical perspective allows the reader to easily and quickly grasp concepts. These goals are important to the author. All material, including the sections covering analytical tools and processes, are presented for immediate application so as to allow the reader to gain appreciation on the use of these concepts and how they can be implemented in their own organizations.

This book is recommended for those readers who are in the quality and reliability profession and for management personnel of organizations seeking the opportunity to either develop or improve their product design or process quality. In addition, the book is an excellent overview of product assurance planning, quality and reliability tool application, training, and cost of quality management for students in engineering, statistics, and engineering management programs. It should be noted that this book would be further valued by the reader after he or she has received basic exposure to classical methods of quality and reliability. This introduction to the basic methods of quality and reliability would help prepare the practitioner for the product assurance principles necessary in managing an effective product assurance organization.

The material in this book is meant to help people in technical business environments develop a product assurance activity and effectively impact the quality and reliability of the particular product being designed and manufactured. It contains a discussion of hardware as well as some software assurance tools. Many of these tools and processes have been extensively used by quality and reliability professionals. However, the problem has always been how these particular tools and tasks are used to effectively impact the improvement of the product quality and reliability throughout the product's development cycle. This publication helps to inform the reader on the principles and application of product assurance within the organization. It focuses on the what, why, when, where, who, and how of product assurance implementation and management.

The author hopes this material allows a significant understanding and appreciation of product assurance principles and application. In addition, this book can be a key reference to all practitioners of product assurance in their organizations.

ACKNOWLEDGEMENT

For their inspiration, support, and consultation I would like to thank my father, John P. Bieda, and mother, Jeanette Bieda.

John Bieda, PE, CQE, CRE

Introduction

Product assurance is an activity which some companies have but only a few properly implement. The assurance activity cannot be used only as a symbol of intent toward quality and reliability improvement, but rather, it must be pursued with total commitment. Product assurance must take a dedicated commitment from top management on down to the various departments of an organization. Only through the efforts of senior management can the elements of product assurance significantly impact a product's quality and reliability.

Many years of experience in the field of product assurance has allowed the author to understand how product assurance is implemented and properly managed in order to gain the most return on investment—high quality and reliability at the least cost and in the shortest time. It is the use of product assurance planning, strategic incorporation of product assurance tools and processes, just-in-time training, and the application of cost of quality management which help to effectively impact the design and manufacture of products. This product assurance approach can be applied to all phases of a product's development cycle and can be used to affect the design, manufacture, and assembly of a new or modified component, subsystem, or system. The real challenge to a business is how to effectively implement product assurance within its own organization.

In order to accomplish the task of preparing the reader to implement and manage product assurance in his or her organization, the material presented must be practical and efficiently organized so that all portions of the product assurance process are understood and can be applied to any phase of a product development cycle. First, the book will expose the reader to the principal background material of product assurance and help answer important questions of what, why, when, where, who, and how. Second, past and present perspectives of product assurance in terms of mass versus lean production, Deming principles and their application, and the relationship of QS 9000 requirements to product assurance will be reviewed to illustrate the progress and focus of the product assurance profession. Third, the book will begin discussing how the product assurance plan (PAP) is developed and relates to the product development process (PDP) of any component or system. Following the review of the PAP, an extensive discussion will be made regarding the application of specific hardware and software tools and processes which help to assess and improve product designs and manufacturing processes. The next chapters of the book will talk about product assurance training, cost of quality management, and managerial practices. Finally, the remaining chapters will discuss future direction of product assurance and a summary of its benefits.

List of Tables

Table Number	Title	Page
Table 4a	Bayesian Quantification of Past Experience	52
Table 4b	Reliability Demonstration Sample Size Requirements (Based on Bayesian Techniques)	53
Table 5a	Average and Range Method Constants	96
Table 5b	Reliability Growth Model Parameters—Electromechanical Device Example	107
Table 5c	Reliability Growth Test Plot Data—Electromechanical Device	107
Table 6a	Abbreviated Table of Chi-Square Values	119
Table 6b	Table of Critical Values of D in the Kolmogorov-Smirnov Goodness of Fit Test	123
Table 6c	Abbreviated Table of F Critical Values for a Two-Tailed Test of Variance	125
Table 6d	Abbreviated Table of t Critical Values for a Two-Tailed Test of Means	126
Table 7a	Table of Constants and Formulas for Control Charts	179

List of Figures

Figure Number	Title	Page
Figure 1.1	Definition of Product Assurance	2
Figure 1.2a	Philosophy of Operation	3
Figure 1.2b	Quality System Documentation Progression	4
Figure 1.3	Hardware versus Software Reliability	4
Figure 1.4	Hardware and Software Assurance Relationship to PDP	5
Figure 1.5	Reasons for Product Assurance Management	6
Figure 1.6	When Product Assurance Is Implemented	7
Figure 1.7	Product Assurance Effectiveness during Product Development	7
Figure 1.8	Product Design/Process Development Status and Product Assurance Involvement	8
Figure 1.9	Product Complexity Level and Product Assurance Involvement	8
Figure 1.10	Relationship of Product Assurance Activity to Company Organization	9
Figure 1.11	How Product Assurance Management Is Implemented (Order of Execution)	10
Figure 1.12	Implementation of Product Assurance Methods and Tools	11
Figure 1.13	Steps in Forming a Product Assurance Organization	12
Figure 2.1	A Comparison of Assembly Plant Performance Characteristics (Averages of Assembly Plants in Each Region, 1989)	14
Figure 2.2	Comparison of Automotive Product Development Performance (Mid-1980s)	15
Figure 2.3	Relationship of Product Assurance Management Principles and Lean Production	17
Figure 2.4	Relationship of Product Assurance Management Principles and the Deming 14 Points	19
Figure 2.5	Relationship of Product Assurance Management Principles and QS-9000 Quality System Element	21
Figure 3.1	Product Development Process and Major Engineering Milestones	25
Figure 3.2	Relationship of Product Assurance Method to PAP	26
Figure 3.3	Product Assurance Plan Worksheet—Example	28
Figure 4.1	Traditional Four QFD Phases	34
Figure 4.2	Quality Function Deployment Technique ("House of Quality")	35
Figure 4.3a	Quality Function Deployment Example	36
Figure 4.3b	Quality Function Deployment Example	37
Figure 4.4	Benchmarking Process Steps	39
Figure 4.5	Benchmarking Example	40
Figure 4.6	Benchmarking Analysis Matrix Example	43
Figure 4.7	Test Development Conditions and Requirements—Example	45
Figure 4.8	Life Test Profile (Combined Power and Temperature)—Example	46

Figure Number	Title	Page
Figure 4.9	Verification Test Sequence (Series/Parallel Arrangement)—Example	47
Figure 4.10	Test Summary Report Switch Assembly—Example	48
Figure 4.11	Test Sample Planning Examples	50
Figure 4.12	Minimum Reliablility Determination for a 90% Confidence Level from the Number of Failures and Sample Size	52
Figure 4.13	Sample Size Determination from Reliability and Confidence Level (Binomial Distribution Approximation)	53
Figure 4.14	Binomial Distribution Nomograph	54
Figure 4.15	Four Phases of Software Development	56
Figure 4.16	Software Reliability Test Development and Analysis Sequence	57
Figure 4.17	Zone of Operation for Software Inputs	58
Figure 4.18a	Cumulative Errors vs. Test Iteration—Example	61
Figure 4.18b	Probability of Success vs. Test Iteration—Example	61
Figure 5.1	Functional Block Diagram—DC Motor (Motor) Example	65
Figure 5.2	Reliability Block Diagram—A DC Motor Example	66
Figure 5.3	Process Flowchart—An Electronic Module Example	67
Figure 5.4a	Design Failure Mode and Effects Analysis (DFMEA)—Example	69
Figure 5.4b	Process Failure Mode and Effects Analysis (PFMEA)—Example	71
Figure 5.5	DFMEA and Its Relationship to Other Design Development Activities	72
Figure 5.6	PFMEA and Its Relationship to Other Process Development Activities	72
Figure 5.7a	Fault Tree Analysis—A DC Motor Assembly (Top-of-Tree Diagram) Example	74
Figure 5.7b	Fault Tree Analysis—A DC Motor Assembly (Armature Fails to Turn Sub-Tree) Example	75
Figure 5.8	Fault Tree Analysis Fussel-Vesely Importance Measures—A DC Motor Assembly Example	76
Figure 5.9	Descriptive Model for DFMA Evaluation Process	77
Figure 5.10	Design for Manufacturing and Assembly—Guidelines to Good Design	78
Figure 5.11	Design for Manufacturablility and Assembly—Example	79
Figure 5.12	Variation Simulation Analysis Philosophy—An Example	80
Figure 5.13	VSA Evaluation Process Procedure	81
Figure 5.14a	Variation Simulation Analysis Example—Widget Assembly Drawing	82
Figure 5.14b	Variation Simulation Analysis Example—Mathematical Model and Input/Output Variable Specification File	82
Figure 5.14c	Variation Simulation Analysis Example—Simulation Results for Original Condition	83
Figure 5.14d	Variation Simulation Analysis Example—Simulation Results for Original Condition	84
Figure 5.14e	Variation Simulation Analysis Example—Simulation Results for Revised Condition	85
Figure 5.14f	Variation Simulation Analysis Example—Simulation Results for Revised Condition	86
Figure 5.15	Process Control Plan Relationship to Other Process Development Activities	87
Figure 5.16	Process Control Plan—Electronic Circuit Board Manufacturing Process Example	88
Figure 5.17	Measurement System Calibration Analysis—Measurement Accuracy Example	92

Figure Number	*Title*	*Page*
Figure 5.18	Variable Measurement Systems Repeatability and Reproducibility Study—Range Method (Short Method) Example	94
Figure 5.19	Variable Measurement Systems Repeatability and Reproducibility Study—Average and Range Method (Long Method) Example	95
Figure 5.20	Attribute Measurement Systems Repeatability and Reproducibility Study—An Example	97
Figure 5.21	Maintainability Study Example	99
Figure 5.22	Maintainability versus Time to Repair—An Example	101
Figure 5.23	Scheduled Preventive Maintenance Plan—Printed Circuit Board Assembly—Example	102
Figure 5.24	Reliability Growth Test Management Process	104
Figure 5.25	Reliability Growth Test Management Example	106
Figure 5.26a	Reliability Growth Test Management—MTBF Distribution Profile	108
Figure 5.26b	Reliability Growth Test Management—Reliability Profile Example	108
Figure 5.27	Proces Reliability Growth Management Example	109
Figure 5.28	Design Reviews and Their Relationship to Product Development	110
Figure 6.1	Variable Test Data Analysis and Root Cause Spreadsheet—Example	115
Figure 6.2a	Chi-square Test for Normality Example	117
Figure 6.2b	Chi-square Test for Normality Example (Continued)	118
Figure 6.3a	Kolmogorov-Smirnov (KS) Test for Normality Example	119
Figure 6.3b	Kolmogorov-Smirnov (KS) Test for Normality Example	121
Figure 6.3c	Kolmogorov-Smirnov (KS) Test for Normality Example	122
Figure 6.4	Tests of Significance	125
Figure 6.5	Attribute Data Analysis Examples	129
Figure 6.6	Stress/Strength Analysis Technique—Example	133
Figure 6.7	Reliability Improvement through Stress/Strength Analysis	133
Figure 6.8	Reliability Test Analysis Examples	135
Figure 6.9	Reliability Bathtub Curve and Its Implications	139
Figure 6.10	Weibull Distribution Analysis Example	140
Figure 6.11a	Weibull Graph Data—Electromechanical Switch Assembly Life Test Example	141
Figure 6.11b	Weibull Graph—Electromechanical Switch Assembly Life Test Example	141
Figure 6.12	Basic Types of Reliability Block Diagrams and Their Mathematical Representations	144
Figure 6.13	System Reliability Block Diagram Analysis—DC Motor Example	145
Figure 6.14	System Reliability Analysis Summary Chart—DC Motor Example	147
Figure 6.15	Temperature Impact Profiles—DC Motor Example	147
Figure 6.16	An Approach for the Analysis of Variable Data Process Performance or Capability	150
Figure 6.17a	Process Capability Study Based on Determination of Normality—Example	152
Figure 6.17b	Process Capability Study Based on Determination of Normality—Example	153
Figure 6.18	Process Performance, Pp (Indicator of Spread)	154
Figure 6.19	Proces Performance, Ppk (Indicator of Location)	155
Figure 6.20	"On-Target", Pp – Ppk (Indicator of Nominal Location)	156
Figure 6.21	Process Performance Evaluation Example	157
Figure 6.22	Design of Experiment Process Flowchart	159
Figure 6.23	Design of Experiment Analysis Example	160

Figure Number	Title	Page
Figure 6.24	Design of Experiment—Mechanical Spring Assembly Example	161
Figure 6.25	Relationship Between QS-9000 and Proces Review Elements	165
Figure 6.26	Acceptance Sampling	166
Figure 6.27a	Attribute Sampling Plan (MIL-STD-105D) Example	167
Figure 6.27b	Operating Characteristic Curves	168
Figure 6.28	Relationship of Error and Mistake Proofing to Design/Process Development Activities	170
Figure 7.1	Statistical Process Control Examples	175
Figure 7.2	Statistical Process Control Example—X and R Charts	178
Figure 7.3	Statistical Process Control Example—P-Chart	181
Figure 7.4	Pareto Chart Example—Total Field Returns of a Product	183
Figure 7.5	Root Cause Analysis Process	185
Figure 7.6	Root Cause Analysis Example—Sounding Device Failure	186
Figure 7.7	Root Cause Analysis Example—Sounding Device Failure (An Illustrative Representation)	187
Figure 8.1	Types of Short- and Long-Term Training Material	191
Figure 8.2	Qualifications for Product Assurance Trainers and Trainees	193
Figure 9.1	Product Assurance Cost of Quality Management Techniques	196
Figure 9.2	Product Assurance Quality Cost Elements and Their Components	197
Figure 9.3	Cost of Quality Analysis Procedure	197
Figure 9.4	A Simplified Life-Cycle Cost Formulation	198
Figure 9.5	Life-Cycle Cost Analysis Flowchart	199
Figure 9.6a	Life-Cycle Cost Analysis—Example Direct Repair Cost Matrix	199
Figure 9.6b	Life-Cycle Cost Analysis—Example Cost Breakdown Matrix	200
Figure 9.6c	Life-Cycle Cost Analysis—Example Cost Impact Summary Matrix	200
Figure 9.7	Comparison of Total Company Loss and LCC for a Product Operating Period—Example	201
Figure 9.8a	Quality Costs and the Product Development Process (Desired Condition)	202
Figure 9.8b	Quality Costs and the Product Development Process (Actual Condition)	203
Figure 9.9	Cost of Quality Management Example	204
Figure 10.1	Highlights of Preferred Product Assurance Managerial Practices and Behavior	207
Figure 11.1	Product/Process Reliability Development and Improvement Approach	212
Figure 11.2	Reliability Development and Improvement Approach to the Design of a Bicycle Brake System—Example	213
Figure 11.3a	A Comparison of Accelerated Testing Approaches—Constant Stress versus Step Stress	214
Figure 11.3b	Constant and Step Stress Methodologies	215
Figure 11.4	System Engineering Evaluation Process	217
Figure 11.5	Product Assurance Risk Reduction Technique—A Medical Treatment Example	219

Structure of Product Assurance and Its Management

"Transformation is required in government, industry, education. Management is in a stable state. Transformation is required to move out of the present state. The transformation required will be a change of state, a metamorphosis, not mere patchwork of the present system of management. We must of course solve problems and stamp out fires as they occur, but these activities do not change the system."

W. EDWARDS DEMING
"Foundation for Management
of Quality in the Western World"
Deming Management at Work,
1991, Ch. 1, p. 11.

Chapter Introduction

The purpose of this chapter is to discuss the basic concept of product assurance in terms of: (1) what it is, (2) why do it, (3) when should it be implemented, (4) where it is applied, (5) who does it, and (6) how it is managed and implemented. The relationship between product assurance philosophy and its application to the product development process is an important idea discussed in this chapter. Further, this chapter will help introduce the practitioner to the key elements of product assurance planning, quality and reliability tool application in the product development process, training, and cost of quality management.

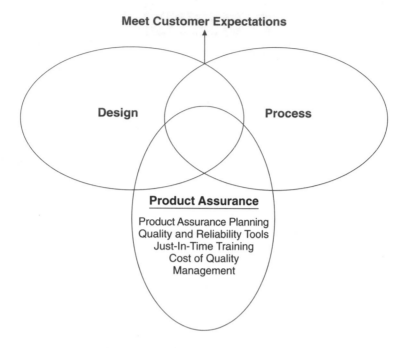

Figure 1.1   Definition of Product Assurance

A. Definition of Product Assurance

Product assurance (PA) is an activity performed by an organization which works with product design and process engineering to meet customer requirements and "build-in" quality and reliability early during the product's development cycle. The assurance objective is met when the product meets specific requirements on the fit, form, function, manufacturability, and assembly. One key element of product assurance, which makes it most effective in product development, is the timely incorporation of particular quality and reliability (Q and R) engineering tools into traditional disciplines such as electrical and mechanical engineering in order to produce a robust design and process. Other elements of product assurance that facilitate management of the discipline involve product assurance planning (PAP), just-in-time (JIT) training, and cost of quality management. All of these elements constitute a product assurance function that is fully integrated into the product design and process engineering groups of an organization. The product assurance group along with design and process engineering personnel provide the simultaneous engineering effort necessary to help meet customer expectations (Figure 1.1).

The product assurance activity in the organization also helps to bridge the gap between QS-9000 requirements and product design/process development (see Figures 1.2a and 1.2b). The specific elements included in the QS-9000 that help to meet the quality and reliability expectations of the customer are addressed through the implementation of primary elements of product assurance management. The objective for Quality System Requirements (QS-9000) is the development of fundamental quality systems in an organization that provide for continuous improvement with emphasis on defect prevention and reduction in variation. ISO 9001: 1994 Section 4 has been adopted as the foundation for QS-9000. Implementing the four principal PA elements helps to address the specific Q and R issues posed in QS-9000 while maintaining support through the product development process. Each element of QS-9000 is addressed in a timely manner through

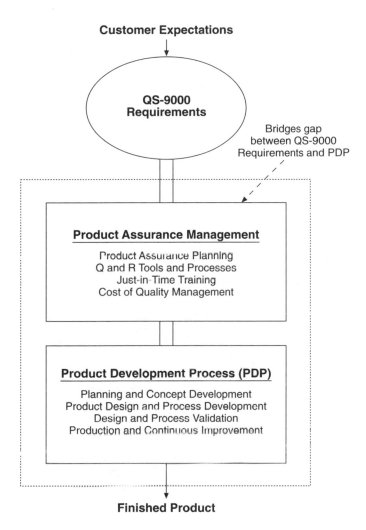

Figure 1.2a Philosophy of Operation

the framework of the PDP and the strategic use of PA management techniques such as PAP, Q and R tool application and consultation, JIT training, and cost of quality management.

Product assurance can be divided into hardware and software assurance. Hardware assurance involves the ability to incorporate those elements of product assurance such as planning, quality and reliability analysis, training, and cost of quality management to impact product design and process development for those components that include certain elements of traditional engineering disciplines. Software assurance involves similar elements as those for hardware assurance, but specifically applies to logic code necessary to run and control hardware functions. Both types of product assurance involve similar tasks but may vary on the approach used to evaluate the reliability of the product (Figure 1.3). The most significant difference in the kinds of Q and R tools used for software versus hardware assurance appears in test sample planning and analysis (see chapter 4, section C, parts 1 and 2). Hardware and software assurance occur simultaneously during product development and affect customer expectations for high quality and reliability, minimum cost, and short development time (see Figure 1.4).

Source: Chrysler, Ford, and General Motors Supplier Quality Requirements Task Force, Quality Systems Requirements QS-9000, 1994, p. 3. Reprinted with permission from the *QS-9000* (Chrysler, Ford, General Motors Supplier Quality Requirements Task Force).

Figure 1.2b   Quality System Documentation Progression

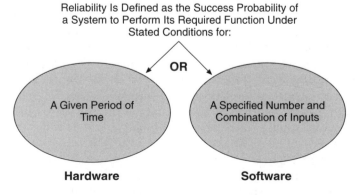

Figure 1.3   Hardware versus Software Reliability

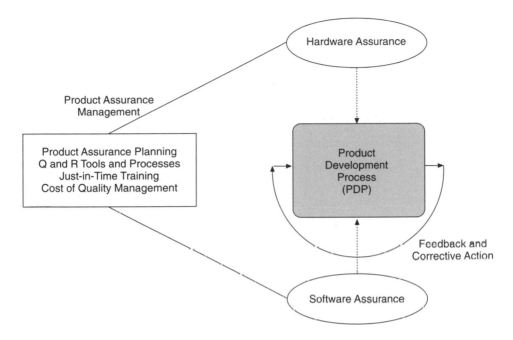

Figure 1.4 Hardware and Software Assurance Relationship to PDP

B. Reasons for Product Assurance

There are several reasons for using product assurance in an organization. In general, the combined use of product assurance planning, quality and reliability tool application, just-in-time training, and cost of quality management helps to satisfy and exceed customer expectations for: (1) high quality and reliability, (2) minimum cost, (3) reduced development time, (4) interest for new product feature and function, (5) improved safety, and (6) meeting environment standards. Product assurance can be viewed as the force that "levers" customer expectations using the PDP as the fulcrum (Figure 1.5).

Accomplishing each of these customer requirements involves one to several of the PA activities. First, meeting high quality and reliability in a reduced development period requires strategic *product assurance planning*. This planning helps to define the specific period in the product development cycle where various *quality and reliability tools* are applied to best impact the design and process of hardware or software and achieve important product development milestones. The use of quality and reliability tools helps to better understand a design or process and facilitates the simplification of the design for improved quality and reliability and the reduction in cost. The PAP helps to shift tasks to earlier in the PDP where prevention of defects can be applied more effectively. *Just-in-time training* helps product design and process engineering be better aware of the application of those quality and reliability tools that could help to improve their designs and minimize extensive testing. The customer expectation for new product feature and function is another requirement in the product planning activity. To satisfy this expectation, product assurance offers the flexibility through its product assurance plan to accommodate the tasks necessary to meet diverse engineering development milestone objectives. The application of *cost of quality management* helps to assess the result of Q and R tool implementation and direct management to the activities that are more prevention-oriented as opposed to reaction-oriented. Reaction-oriented activities involve excessive inspection and test. Last of all, specific PA quality and reliability tools help to determine the risk attached in meeting safety and environmental requirements for a product.

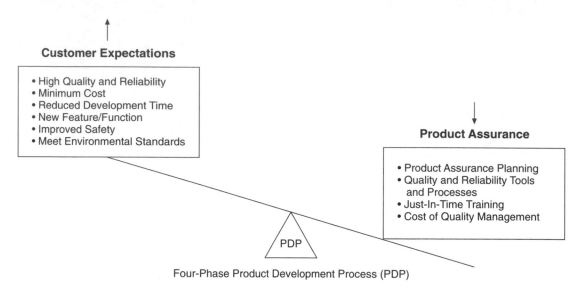

Figure 1.5 Reasons for Product Assurance Management

Each product design and process requires an understanding of how robust each design or process operation is to various environmental stresses and the ability to meet changing customer expectations. Product assurance helps to address these issues by evaluating risk and attempting to minimize the risk of failure in the design and process. This is achieved through frequent evaluations of risk during simulation and actual testing. Next, product assurance is important for the use of various quality and reliability tools throughout the product's development cycle. The strategic use of these tools to "build-in" the quality and reliability is a significant asset of product assurance. The other key attribute of product assurance is the planning of various activities leading to key milestones of a product's development cycle.

C. When Product Assurance Is Implemented

Product assurance is used periodically throughout each phase of a component or system's product development cycle (Figure 1.6). This cycle is divided into four specific phases: (1) planning and concept development, (2) design and process development, (3) design and process validation, and (4) production and continuous improvement. Each phase includes the incorporation of key elements of practical product assurance management, which are product assurance planning, quality and reliability tool application, just-in-time training, and cost of quality management. The use of specific analysis tools helps to assess, design, and improve a product's design and manufacturing process. These product assurance activities provide effective impact when they are applied early in the product development cycle, particularly during the design and process development phase (Figure 1.7). The product assurance function helps to shift the traditional behavior of producing engineering changes late in the development cycle to early phases of product design and process development. It is most economical to incorporate engineering change and reevaluate the design or process during the design and process development cycle, rather than later during process verification or production, when changes are very expensive.

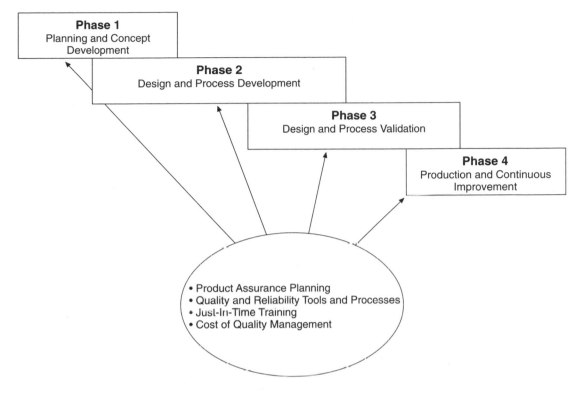

Figure 1.6 When Product Assurance Is Implemented

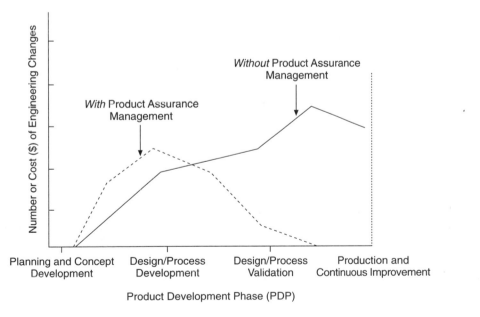

Figure 1.7 Product Assurance Effectiveness during Product Development

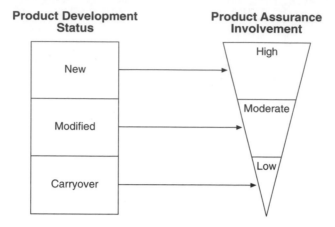

Figure 1.8 Product Design/Process Development Status and Product Assurance Involvement

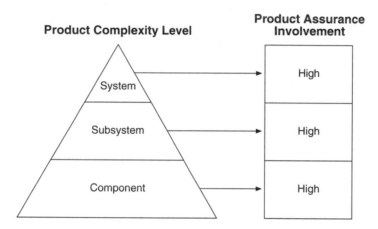

Figure 1.9 Product Complexity Level and Product Assurance Involvement

D. Where Product Assurance Is Applied

Product assurance is applied to new, modified, or carryover product (see Figure 1.8). The use of PA should be considered for each component as the product is defined by the customer at the start of a program. In order to establish the degree of product assurance support, the design and process should be carefully reviewed before the appropriate tasks in the PAP are identified. If the product is new, then there are several PA tasks involved to accommodate the full development process. However, if the product is modified or is a carryover product, then the PAP should be adjusted accordingly to form the PA strategy necessary to accomplish engineering milestones. Product assurance is used to evaluate not only the product's design but the process used to manufacture and assemble the product.

Product assurance can be applied for any product complexity level—component, subsystem, and system, and at all levels of a product's development, even if the organization is responsible only for the design or assembly (Figure 1.9). The level of product design or process complexity should not inhibit use of PA activity, since at any level of a design or process there is a potential for failure and, more important of all, customer dissatisfaction. Regardless of the level of product complexity there should be a PAP with specific PA tasks that involve the use of quality and relia-bility tools. The degree of analysis and test will be based on the complexity and part status. If a system-level PAP is pursued, then the subsystem and component-level PAP should also be pursued

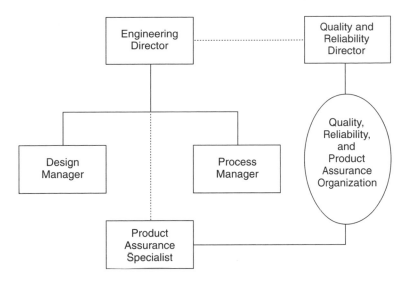

Figure 1.10 Relationship of Product Assurance Activity to Company Organization

in order to assure that the elements of the component or subsystems do not generate sufficient risk to affect the system quality and reliability. If the component or subsystem is designed or manufactured at a different supplier, then the PAP process should be implemented according to the requirements set forth by the primary manufacturer.

E. Organizational Structure of Product Assurance

Product assurance can be organized in such a way as to support an engineering function and act as an independent department (see Figure 1.10). These engineering functions involve design and manufacturing. Each department would be led by a manager who would report to a director of engineering for the particular part of the subsystem. The PA group would consist of several hardware and software engineering specialists who would work alongside the respective product engineers in the specified engineering departments. These specialists would help implement the PA process throughout the development of the product. They would be working with managers from design, manufacturing, purchasing, and quality engineering for the purposes of conducting PA activities and providing support toward important decisions during the product development. This organization of PA personnel helps to promote the simultaneous engineering philosophy among the groups, which is a primary element for support of the lean production environment. The PA group for hardware and software specialists would report to a PA manager, who would then report to the director of engineering.

F. Management and Implementation of Product Assurance

Effective product assurance management involves identifying and implementing the tasks necessary to thoroughly develop the product design and process. It specifically involves positioning the PA hardware or software specialist to incorporate the appropriate quality and reliability tool to augment the design and process development activity. The most effective management of product assurance occurs when the application of its principles are performed in a logical manner during the product development cycle (see Figure 1.11). The PAP initiates the planning of Q and R activities, JIT training reinforces the application, and cost of quality management assesses the effectiveness of the type of Q and R tool used to improve the product.

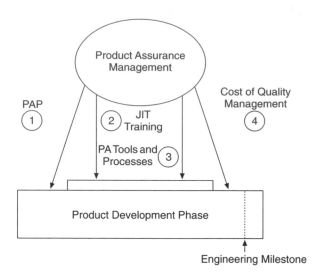

Figure 1.11   How Product Assurance Management Is Implemented (Order of Execution)

The selection of the type of tool and when to use the most appropriate one for maximum benefit is an important element for managing the product assurance activity (see Figure 1.12). The use of the PAP for scheduling where particular tools should be used is important to assess product integrity to the intended requirements. It doesn't make sense to spend a significant amount of time and effort in the application of a quality or reliability tool when it has been determined that the tool doesn't provide any value to the product's ability to meet customer's expectations. Therefore, significant attention of PA management is to match the benefit to the kinds of tools that would help to properly evaluate the design and promote improvement.

Appropriate training is needed for the engineering staff and suppliers on specific processes implemented in the PAP, which would help integrate various tools and processes without the constant assistance of a PA engineer. Training is needed to reinforce PA activities necessary for first-tier suppliers who interface with the customer. These training sessions may involve areas such as concept selection criteria, accelerated testing, product design and process verification, process review, field return and analysis programs, etc. This training should be performed in a just-in-time fashion in order to properly prepare engineering or the supplier community for activities that will be used in the immediate future.

The primary tools and processes used to accomplish engineering milestone events should be organized to support the other. Those tools used in the concept development phase should facilitate the development of engineering requirements and produce design and process definition. The processes used in the design development phase should include simulation analyses, test data analysis, risk evaluation, and design for assembly techniques. Design reviews and the use of the failure modes and effects analysis (FMEA) and fault tree analysis (FTA) should be used throughout to help collect information, reassess risk, and formulate conclusions from the application of these techniques on design and process integrity. Next, there should be a relationship between the process flowchart, process failure modes and effects analysis (PFMEA), and process control plan in helping to establish concise operating instructions for the manufacture of the product. Measurement analyses followed by capability studies are performed to evaluate the ability of the various machines and operations to perform assembly or manufacturing operations in a consistent way without potential for defects. The use of error and mistake-proofing techniques, preventive maintenance actions, and process controls helps to affect the cost of quality. Finally, the use of various techniques such as the Pareto chart, design of experiments, and root cause analysis helps to

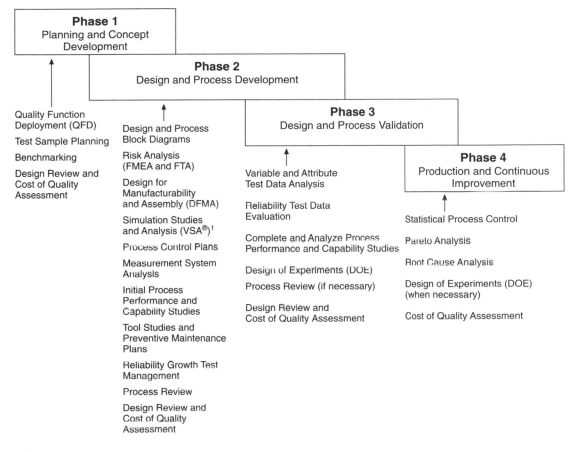

Figure 1.12   Implementation of Product Assurance Methods and Tools

direct continuous improvement of the design and manufacture of the product. In summary, it is important that the choice of PA tools and processes compliment one another and collectively address questions of the product design and process.

The implementation of the last important element of practical product assurance management is the cost of quality. PA should account for how the various techniques and processes of PA evaluation are affecting the balance of prevention, appraisal, internal, and external failure costs. This activity should resume through production in order to compare product performance in the field to development expenses.

The steps that can be applied to establish a product assurance organization and implement their principles are summarized in Figure 1.13. The following list is an example of these steps:

1. Obtain senior management's understanding and full commitment to the product assurance activity. Explain benefits of PA in building Q and R up front in product development cycle as opposed to traditional reactionary approach after the defect in a design or process has occurred. Explain the relationship of the various tools to product development phase (PAP). Discuss the relationships of Deming principles and the QS-9000 requirements to the product assurance activity.

2. Obtain product assurance organization by acquiring product design and process specialists in hardware and software quality and reliability. Develop organization around commodity areas such as electrical, mechanical, or software embedded components.

3. Understand product development process and identify key engineering milestone events and dates.

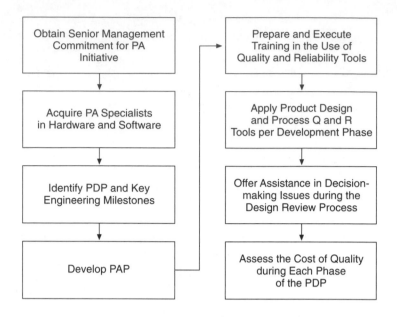

Figure 1.13 Steps in Forming a Product Assurance Organization

4. Develop a simple PAP from the PDP and implement those Q and R tools prior to risk assessment and design reviews for each development phase.
5. Prepare and execute training sessions on particular Q and R tools to be used during product development to help address risk and build-in reliability and quality.
6. Apply product design and process Q and R tools per development phase to assess risk and design in reliability and quality. Approach the evaluation and product design-process development through the use of Q and R tools in series. In other words, build on the results obtained from one tool from the results of another (that is, process flowchart to process FMEA to process control plan, or simulation study to statistical evaluation to product design accelerated life test analysis, etc.). Reflect the results of any analysis to the FMEA. This document should contain the corrective actions and assessment of risk as results of an action done to help detect or prevent a failure condition.
7. Offer assistance in decision making regarding product development impacting quality and reliability. Provide consultation and support to decision-making process for product design and manufacturing.
8. Assess the cost of quality during each phase of product development by benchmarking PA effectiveness on quality cost and life cycle costs. Determine the margin of improvement in the total cost of quality as a function of PA activity throughout the PDP.

NOTES

1. VSA® is a registered trademark of Variation Systems Analysis, St. Clair Shores, Michigan.

2

Past and Present Perspectives on Product Assurance

"Perhaps the most striking difference between mass production and lean production lies in their ultimate objectives. Mass producers set a limited goal for themselves—'good enough,' which translates into an acceptable number of defects, a maximum acceptable level of inventories, a narrow range of standardized products. To do better, they argue, would cost too much or exceed inherent human capabilities.

Lean producers, on the other hand, set their sights explicitly on perfection: continually declining costs, zero defects, zero inventories, and endless product variety. Of course, no lean producer has ever reached this promised land—and perhaps none ever will, but the endless quest for perfection continues to generate surprising twists."

J. P. WOMACK, DANIEL T. JONES, AND DANIEL ROOS,
The Machine that Changed the World, 1990, PP. 13–14.

Chapter Introduction

The purpose of this chapter is twofold: (1) to compare and contrast the past philosophy of mass production to the present philosophy of lean production and (2) to relate the principal elements of practical product assurance to the key elements of lean production and Dr. Deming's 14 Points on quality. This chapter will help to form and enhance the relationship of practical product assurance management to the popular philosophies and strategies in the quality discipline.

Characteristic	Japanese Manufacturer in Japan	Japanese Manufacturer in North America	American Manufacturer in North America
Performance:			
Productivity (hours/vehicle)	16.8	21.2	25.1
Quality (assembly defects/100 veh.)	60	65	82.3
Layout:			
Space (sq. ft./vehicle/year)	5.7	9.1	7.8
Size of Repair Area (as % of assembly space)	4.1	4.9	12.9
Inventories (days for 8 sample parts)	0.2	1.6	2.9
Work Force:			
% of Work Force in Teams	69.3	71.3	17.3
Job Rotation (0 = none, 4 = frequent)	3	2.7	0.9
Suggestions/Employee	61.6	1.4	0.4
Number of Job Classes	11.9	8.7	67.1
Training of New Production Workers (hours)	380.3	370	46.4
Absenteeism	5	4.8	11.7
Automation:			
Welding (% of direct steps)	86.2	85	76.2
Painting (% of direct steps)	54.6	40.7	33.6
Assembly (% of direct steps)	1.7	1.1	1.2

Figure 2.1   A Comparison of Assembly Plant Performance Characteristics[1] (Averages of Assembly Plants in Each Region, 1989)

A. Mass Production versus Lean Production Philosophy

The purpose in this chapter is to discuss the differences in mass production and lean production and how the role of product assurance fits into the lean production philosophy and objectives. Both the mass and lean production systems are significant philosophies, which impact the operation of the manufacturing environment and continually guide the industry into the future. It is worth the time to understand the characteristics of both philosophies and how product assurance management plays a role in achieving the objectives as established in lean production.

The current direction of industry is to move toward a lean production system as opposed to a mass production philosophy of operation, which was prevalent in the past for manufacturers. Mass production was a form of operation first conceived by Henry Ford in 1913 through the production of the Model T Ford automobile and involved a complete and consistent interchangeability of parts and simplicity of attaching them together. This system of mass production continued successfully through many years in the United States until the appearance of lean production by the Japanese in the 1960s. Lean production, as opposed to mass production, advocated teamwork, communication, efficient use of resources, and a continuous improvement philosophy. The result of the lean production approach on the manufacturing industry was: (1) reduction of human effort in the factory, (2) decrease in the investment of tools and fixtures, (3) minimization of manufacturing space, (4) reduction in engineering development time, and (5) decrease in the number of engineering hours.[2] A comparison of the characteristics of lean production versus mass production can reviewed in Figures 2.1 and 2.2 (note the differences in the number of defects, inventory levels, labor hours/unit, etc.).

Product Development Characteristics	Japanese Manufacturer	American Manufacturer
Average Engineering Hours per New Car (millions)	1.7	3.1
Average Development Time per New Car (in months)	46.2	60.4
Number of Employees in Project Team	485	903
Number of Body Types per New Car	2.3	1.7
Average Ratio of Shared Parts	18%	38%
Supplier Share of Engineering	51%	14%
Engineering Change Costs as Share of Total Die Cost	10%–20%	30%–50%
Ratio of Delayed Products	1 in 6	1 in 2
Die Development Time (months)	13.8	25
Prototype Lead Time (months)	6.2	12.4
Time from Production Start to First Sale (months)	1	4
Return to Normal Productivity After New Model (months)	4	5
Return to Normal Quality After New Model (months)	1.4	11

Figure 2.2   Comparison of Automotive Product Development Performance[3] (Mid-1980s)

Mass production philosophy has endured through its development in the early part of the twentieth century. The characteristics of this philosophy can be summarized as follows:[4]

1. Contained single task-oriented manufacturing operations with a high separation of labor. The foreman, industrial engineer or so-called quality engineer, were the only individuals responsible for quality.
2. Mass production included a vertically integrated manufacturing operation. The construction of the product evolved in-house from the basic raw materials to the component and subsystem level.
3. Everything was designed and manufactured in-house versus through the "invisible hand" approach. The "invisible hand" philosophy meant dealing and buying from outside suppliers. These transactions were based on price, delivery, quality, and expectation of the long-term relationship. The in-house approach of doing business was very bureaucratic.
4. A significant number of shipping problems and trade barriers were predominant under mass production.
5. A dependence on a standard product line, including parts and processes was common. An inability to adapt flexible assembly systems or design parts that are favorable to manufacturing capabilities. This approach was not suited for world market conditions.
6. Tools and fixtures were highly specialized, inflexible, and very expensive to change for a new task. Tools were accurate due to the specialization, but not maintained as frequently.
7. If part didn't fit properly or was installed slightly out of tolerance, the owner was expected to fix it.
8. Mass production design and manufacturing processes produced low quality parts, required extensive rework, and frequent end-of-line inspections.
9. Decision-making capability was highly centralized. Make or buy mentality was the only technique commonly practiced.
10. Product development process was sequential and did not allow for any simultaneous development activity.

Lean production mentality was a significant shift from the traditional mass production thinking as practiced by most U.S. manufacturers. This new philosophy contained many of the ideas first conceived by Deming during his work with the Japanese shortly after World War II. Eiji Toyoda, after learning from Deming and observations of the operations at Ford's River Rouge plant in Detroit during the 1950s, instituted and practiced the principles of lean production. These characteristics are summarized as follows:[5]

1. Teamwork was very apparent between the various disciplines within engineering, between departments, and with suppliers. People at all levels of the company worked with one another in agreeing on cost structures, improving quality and reliability, and developing designs that consider requirements for design and manufacturability of the product.

2. There was improved communication across all disciplines of the organization. This enhanced appreciation and confidence in the advancement of the product through its development cycle. Such communication facilitates simultaneous engineering of design and manufacturability of the product, thus decreasing development time and decreasing the amount of defects experienced. The company behaves more like one community rather than separate communities.

3. There was efficient use of resources such as the interchangeability of dies and tool tasks during process development and production.

4. There was a focus on making small batches of product versus producing large amounts in order to eliminate carrying costs and huge inventories. Just-in-time inventory systems were used to accomplish the minimization of the total cost of inventory and the probability of inspection of large lots of material. This helps eliminate the waste incurred by inspection costs and sorting large lots of material for quality problems.

5. Continuous improvement was consistently practiced throughout the product development cycle. The focus was more on making improvements as opposed to responding to problems and making temporary fixes. Problem solving was incorporated at all levels of product design and process development. It was important to trace the root of the problem and then find the fix to prevent future occurrences.

6. Quality and reliability improvement early in the development cycle was a key objective in order to prevent the occurrence of a defect due to the design or manufacture of the product. Shifting the cost of quality to prevention and subsequently to a lower total cost in the long term was the main goal.

7. It was important to work with suppliers at the first- and second-tier levels. The supplier and company worked together throughout product development. A long-term agreement was established for analyzing costs, determining prices, and sharing in the profits. Supplier and company cooperated in improving quality while evaluating cost positions.

8. The evaluation of suppliers was based on the number of defects, delivery performance, and capability in reducing cost.

9. There was a concentration on securing part and subsystem standardization with the ability to incorporate design and manufacturing flexibility.

10. The dealer-to-buyer relationship was customer driven with emphasis on customer satisfaction through repairing problems the first time, providing customer with convenience in service, and issuing customer with incentives to next purchase. The philosophy in customer service was long-term not short-term.

The key role of product assurance management is to incorporate the lean production elements into product design and process development. In other words, product assurance should act as the mechanism during product development to facilitate teamwork, communication, efficient use of resources, and continuous improvement. This facilitation is achieved through incorporation of the product assurance guidelines such as PAP, application of quality and reliability tools, training, and cost of quality management. Each lean production element can be addressed by each

Lean Production Element	Product Assurance Technique				
	Product Assurance Planning	Q and R Tool Implementation	JIT Training	Cost of Quality Management	Management Behavior
Teamwork	X	X	X	X	X
Communication	X	X	X	X	X
Efficient Use of Resources and Elimination of Waste		X		X	X
Continuous Improvement		X	X	X	X

Figure 2.3 Relationship of Product Assurance Management Principles and Lean Production

of the product assurance management guidelines (see Figure 2.3). Specifically how each of these PA guidelines relates to the four lean production elements can be discussed accordingly.

The differences between the mass versus lean producers, which were apparent during the 1980s, have been significantly reduced in the 1990s. The two-to-one difference in most of the product development characteristics during the 1980s, as shown in Figures 2.1 and 2.2, are now almost indiscernible. For example, the number of defects per 100 vehicles has declined significantly for both Japanese and U.S. automobiles, and the gap has narrowed since this study was done (see recent J. D. Powers reports and recent reports by Consumers Union). The incorporation of lean production principles in all phases of the product development process has helped American industries, particularly the auto manufacturers, compete in the worldwide marketplace.

The first element of teamwork in the organization can be correlated to the ability of product assurance to work with each engineering group (design, manufacturing, tooling, materials, etc.) in defining and resolving the design or process opportunity. This activity may involve understanding the design or process opportunity, gathering information, and applying the specific Q and R tool(s) to help detect where appropriate engineering techniques may be applied to improve performance or prevent problems. These PA activities would span through the entire product development process. Another PA management element that can be related to lean production is the use of the PAP. This PA management technique is instrumental in defining which members of the development team play a significant role in accomplishing the activities leading to the engineering milestone within the PDP.

The second element of communication in the organization can be correlated to the fact that product assurance personnel work with the appropriate departments in the organization to gather information and evaluate design and process issues, as well as consult on the most beneficial resolution to the design or process opportunity. The PA engineer must look at each design or process opportunity from the perspective of feasibility, cost, manufacturability, quality, reliability, etc. The PAP helps by identifying the relative time and people responsible for completing the task prior to the key engineering milestone date. Another PA management element that can be related to lean production is just-in-time (JIT) training. The just-in-time training of all people involved in activities in the PDP which affect assessment of quality and reliability is a key communication activity that facilitates awareness and promotes a better understanding of Q and R tool utilization.

The third lean production element concerning the efficient use of resources and elimination of waste can be correlated to PA techniques applied through cost of quality management coupled with utilization of specific Q and R tools. The use by PA of cost of quality management techniques facilitates an understanding of where in the organization there is excessive cost due to failures and inspections. Once the magnitude of the problem is recognized, then the specific use of various Q and R tools or engineering approach can be applied to resolve the concern that led to excessive cost and waste to the organization.

The fourth lean production element of continuous improvement can be directly correlated to the strategic use by product assurance of Q and R techniques and training of the engineering personnel to help recognize the need for a specific technique in detecting and improving a product design or process concern. The cost of quality management tool is very useful in helping measure, in economic terms, the effect that each Q and R tool has on continuous improvement.

In summary, the implementation of product assurance management and its techniques facilitates lean production behavior. Utilization of product assurance techniques during the product development process helps to avoid the old "over the wall" philosophy to product/service design (design engineering designs the product and throws it over the wall to production, who sees it for the first time). A cross-functional team approach, with communication, efficient use of resources, and continuous improvement, are key lean production elements that are facilitated directly by product assurance techniques.

B. Deming's Fourteen Points and Their Relationship to Product Assurance

There is a significant relationship of the principles of practical product assurance management to the 14 points that Deming developed on the subject of how to attain quality in an organization.[6] Understanding this relationship is obtained by simply comparing the four major principles of practical product assurance management to basic groups of the Deming 14 points (see Figure 2.4). Each product assurance element relates to "creating a constancy of purpose," "an adoption of the new philosophy," and to "put everybody to work to accomplish the transformation." The following is a discussion of the relationship of the principles of product assurance management and the Deming points:

1. *Development of a product assurance plan (PAP)*
   The specific Deming points that are addressed by the use of a PAP are: (1) "create a constancy of purpose," (2) "adopt the new philosophy," (3) "put everybody to work to accomplish the transformation," and (4) break down barriers between departments. The PAP is a new philosophy, which accommodates and helps direct the engineering team, along with other departments, to implement various elements of a quality plan throughout a product development cycle. This PAP process helps provide a consistent direction to the team, showing when and where specific quality and reliability tools and processes should be applied. The PAP process helps the engineering team meet quality and reliability objectives prior to meeting key engineering milestones throughout the product development process. It helps to identify the team and allow each member to work with one another to accomplish the goal.

2. *Strategic Implementation of Various Quality and Reliability Tools*
   The particular Deming points that are addressed by strategic implementation of various quality and reliability tools include: (1) "constantly and forever improve the system," and (2) cease dependence on mass inspection to achieve quality. The critical use of various Q and R techniques during the specific phases of the PDP helps to better detect and evaluate the magnitude of variability as compared to tolerance limits present in a product design or process. Early recognition and analysis of the variation in a design or process during development helps to indicate to engineering that performance may be affected and may necessitate closer examination. Variation analysis through the use of various quality and reliability processes enables the engineering team to execute improvements inexpensively prior to production. The execution of several Q and R tools in some logical manner helps to understand the relationship of the product design or process performance to particular product design and process parameters. This logical approach used during the various phases of the PDP allows for continuous improvement, a key element in the lean versus mass production philosophy.

Product Assurance Technique

Deming Quality System Point	Product Assurance Planning	Q and R Tool Implementation	JIT Training	Cost of Quality Management	Management Behavior
Create Constancy of Purpose	X	X	X	X	X
Adopt the New Philosophy	X	X	X	X	X
Cease Dependence on Mass Inspection		X		X	
Constantly and Forever Improve the System		X		X	
Remove Barriers					X
Drive Out Fear					X
Break Down Barriers between Departments	X				X
Eliminate Numerical Goals					X
Eliminate Work Standards					X
Institute Modern Methods of Supervision					X
Institute Modern Methods of Training			X		
Institute a Program of Education and Retraining			X		
End the Practice of Awarding Business on Price Tag				X	
Put Everybody to Work to Accomplish the Transformation	X	X	X	X	X

Figure 2.4   Relationship of Product Assurance Management Principles and the Deming 14 Points

3. *Just-in-Time Training*

   The specific Deming principles that are addressed by just-in-time training include: (1) "institute modern methods of training," and (2) "institute a program of education and retraining." The just-in-time principle of training involves training the team in advance but just prior to the interpretation and/or utilization of a specific Q and R tool, which helps detect, evaluate, or direct attention to risk of a product design or process. Significant comprehension and the success of the application of the tool is highly dependent on immediate implementation on current product design and processes during the specific phase in development. Just-in-time training may be repeated several times to the team until full understanding and appreciation is satisfied. Management attendance and support is needed through actual participation in each program in order to secure the training success.

4. *Cost of Quality Management*

   The particular Deming principles that are covered by cost of quality management include several points such as: (1) "constantly and forever improve the system," (2) "end the practice of awarding business on price tag," and (3) "cease dependence on mass inspection." All of these points have a relationship to the idea of managing the cost of quality by focusing on prevention and the subsequent reduction of the total cost. The idea of improving the system and ceasing dependence on inspection or appraisal activities alludes to the idea of shifting one's efforts to understanding the magnitude and source variation, removing it, and preventing future occurrence of it as opposed to just containing it. Other Deming principles such as ending the practice of awarding the business on price tags is covered in the idea of cost of quality management by focusing where cost can be reduced through improvements to the system that help prevent excessive failure or appraisal costs in the first place. The cost of quality management helps direct the rest of the organization to work toward reduction of variation and implementation of prevention activities.

   Other Deming principles that are addressed by product assurance involve the managerial and behavioral aspects of PA management. The means by which PA management carries out the practical principles of PAP, Q and R tool implementation, just-in-time training, and cost of quality management cannot effectively take place unless a lean production type attitude is applied. For example, those Deming points such as (1) "remove barriers," (2) "drive out fear," and (3) "break down barriers between departments" involve destructive elements of an organization, which impede progress toward understanding where variation is in a product design and process and how to prevent or remove it entirely. A PA organization would enhance the communication and teamwork qualities between various departments in the organization by linking the efforts of design, manufacturing, purchasing, etc., to better understand the product from a customer's perspective. From this perspective these groups see how their combined efforts through PA support and consultation can better address and incorporate performance, manufacturability, quality, and reliability into the product design. Product assurance becomes the "catalyst" for simultaneous engineering toward the customer satisfaction objectives and a valuable asset to the promotion of lean production.

   The remaining Deming points are indirectly covered in the institution of practical product assurance management. The "elimination of numerical goals or work standard" is better replaced by the ability to demonstrate improvement in defect prevention, variation reduction in design and process performance, overall cost of quality. It is the nature behind how the techniques are used to increase the margin of improvement, which is more important than reaching a goal that may or may not be attainable in the time designated and would only disappoint the organization. The last Deming point to be discussed regarding "institution of modern methods of supervision" relates to the integration of various tools and processes in a shorter time period, which helps employees reach better decisions and motivates them to excel and feel good about what they are doing. This will involve better utilization of statistical thinking with other traditional engineering tools in order to address risk assessment for product design and process development.

Quality System Assessment Element	Product Assurance Principle			
	Product Assurance Planning	Q and R Tool Implementation	JIT Training	Cost of Quality Management
Management Responsibility	X	X	X	X
Quality System (Documentation)	X	X	X	X
Contract Review				
Design Control	X	X		
Document and Data Control	X			
Purchasing				X
Control of Customer Supplied Product				
Product Identification and Traceability				
Process Control		X		
Inspection and Testing		X		
Inspection, Measuring, and Test Equipment		X		
Inspection and Test Status		X		
Control of Nonconforming Product		X		
Corrective and Preventive Actions		X		X
Handling, Storage, Packaging, and Delivery	X			
Control of Quality Records	X			
Internal Quality Audits				
Training			X	
Servicing	X			
Statistical Techniques		X		
Production Part Approval Process	X			
Continuous Improvement	X	X	X	X
Manufacturing Capabilities		X		

Figure 2.5 Relationship of Product Assurance Management Principles and QS-9000 Quality System Element

C. QS-9000 Requirements and Their Relationship to Product Assurance

Just as the practical principles of practical product assurance management addressed Deming's 14 points, they also addressed key elements of the QS-9000 requirements. In essence, product assurance management functions as a bridge between the QS-9000 requirements and the product development process (see Figure 1.2). These principles of practical product assurance management help to address most of the QS-9000 elements specifically through the strategic involvement of PA personnel and the utilization of the Q and R tools during each phase of the product development process.[7] The questions posed in the *Quality System Assessment (QSA)* manual are structured for each QS-9000 element.[8] The four basic principles of practical product assurance management can be compared to each element in the QS-9000 (see Figure 2.5). A review of the specific tools and processes that the PA engineer would apply to address each element in the QSA can be found in chapters 4–9.

As can be seen by the matrix in Figure 2.5 most of the elements are specifically addressed by implementation of the various PA principles. Those elements, which include: (1) contract review, (2) purchasing, and (3) document and data control, are exclusively handled by other organizational groups but may consult on product assurance when necessary. The management responsibility element of QS-9000 is shared between the PA management personnel and engineering, but ultimately is handled by the upper management staff of the organization. It is primarily up to senior management to communicate the commitment of the quality system, who is ultimately responsible, and how it is to be enforced. Product assurance management will play the role of linking the requirements and helping to satisfy them during the product development process, but ultimately it is upper management that is responsible for the awareness, commitment, and accountability.

NOTES

1. IMVP World Assembly Plant Survey, 1989, and J. D. Power Initial Quality Survey, 1989. Originally adapted by James P. Womack, Daniel T. Jones, and Daniel Roos, *The Machine That Changed the World.* (New York: Harper Collins Publishers, 1990), p. 92.

2. Womack, James P., Daniel T. Jones, and Daniel Roos, *The Machine That Changed the World.* (New York: Harper Collins Publishers, 1990), p. i.

3. Clark, Kim B., Takahiro Fujimato, and W. Bruce Chew, "Product Development in the World Auto Industry," "Brookings Papers on Economic Activity," No. 3, 1987; and Takahiro Fujimato, "Organizations for Effective Product Development: The Case of the Global Motor Industry," Ph.D. Thesis, Harvard Business School, 1989, Tables 7.1, 7.4, and 7.8. Originally adapted by James P. Womack, Daniel T. Jones, and Daniel Roos, *The Machine That Changed the World.* (New York: Harper Collins Publishers, 1990), p. 118.

4. Womack, James P., Daniel T. Jones, and Daniel Roos, *The Machine That Changed the World.* (New York: Harper Collins Publishers, 1990), pp. 21–47.

5. ———, pp. 48–69.

6. Scherkenbach, William W., *The Deming Route to Quality and Productivity: Roadmaps and Roadblocks.* (Rockville, MD: Mercury Press, 1988).

7. Chrysler, Ford, and General Motors Supplier Quality Requirements Task Force, *Quality System Requirements QS-9000,* 1994.

8. Chrysler, Ford, and General Motors Supplier Quality Requirements Task Force, *Quality System Assessment (QSA),* 1994.

3

Product Assurance Planning Within the Product Development Process

"Everyone can take part in a team. The aim of a team is to improve the input and the output of any stage. A team may well be composed of people from different staff areas. A team has a customer.

Everyone on a team has a chance to contribute ideas, plans, and figures; but anyone may expect to find some of his best ideas submerged by consensus of the team; he may have a chance on the later time around the cycle. A good team has a social memory.

At successive sessions people may tear up what they did in the previous session and make a fresh start with clearer ideas. This is a sign of advancement."

W. EDWARDS DEMING,
Out of the Crisis, 1986, CH. 2, PP. 89–90.

Chapter Introduction

The purpose of this chapter is to introduce the first of four key elements of practical product assurance management: product assurance planning (PAP) within the product development process. The primary objective of product assurance planning is to implement a project management tool for utilizing the appropriate quality and reliability tools (addressing what, where, when, and who) necessary in facilitating the product development process. An organized schedule is important for strategically implementing appropriate quality and reliability tools.

A. Product Development Process (PDP)

An efficient product development process has become a necessity in the design and manufacture of a product, particularly when competitive market forces have caused a demand for accelerated development time, reduced cost, and increased quality and reliability. To significantly decrease the time required to develop new or modified products, while pursuing activities that meet customer requirements and enhance customer satisfaction, involves a progressive and simultaneous set of engineering events. The concept of *simultaneous engineering,* with its concurrent product design and process development and validation activities, has been effective in reducing development time and improving teamwork and communication in the organization. In addition, the use and relationship of the various engineering tools and processes throughout the simultaneous engineering process is vital in order to significantly impact the quality and reliability of the product.

The simultaneous engineering process can also be referred to as the product development process (PDP). It can be organized into four overlapping phases (see Figure 3.1):

1. Planning and Concept Development
2. Design and Process Development
3. Design and Process Validation
4. Production and Continuous Improvement

These overlapping phases help prove to be an effective means for the reduction in product development time and the promotion of improved communication and teamwork. Further, the overlapping of the design and process development phases reduces development time by allowing activities to occur at the same time as opposed to being in series. The achievement of improved quality and reliability comes through the careful integration and implementation of various quality and reliability engineering tools and processes. The strategic use of specific tools and processes and their incorporation into the PDP help "build-in" quality and reliability while shifting defect prevention to the early phase of product design and process development.

The product assurance organization is the key contributor to the engineering community, especially when it comes to identifying when and what type of quality and reliability engineering tool to incorporate during the product development process. Product assurance management and engineering events and management must work as one complete organization to assure that each of the engineering events and activities during product development are addressed and implemented. The objective of product assurance planning (PAP) is to properly integrate the tool or process into the PDP.

B. Product Assurance Planning (PAP)

Product assurance planning is a unified approach for integrating the areas of quality, reliability, maintainability, and serviceability into the product development process. It is through product assurance planning that the requirements for quality, reliability, and customer satisfaction are met.

Product assurance planning is an important activity to help identify the "building blocks" of quality and reliability and to improve the design and manufacture of the product. The PAP is a prevention-oriented strategy used consistently throughout the product development process in facilitating design and manufacture of a product that meets and exceeds the customer's expectation. This PAP process involves individuals from various parts of the company's organization such as product planning, product design, engineering, manufacturing, quality and reliability, assembly, purchasing, finance, sales and marketing, suppliers, and other departments.

A successful PAP is dependent upon senior management's commitment to the product assurance process and effort required to achieve customer satisfaction. The effectiveness of the product development process to achieve customer satisfaction cannot be achieved until management has

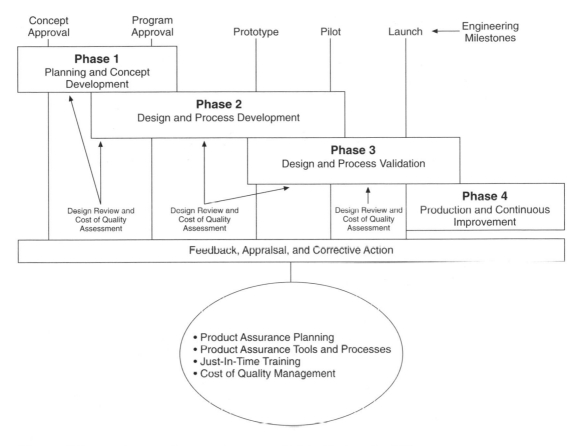

Figure 3.1 Product Development Process and Major Engineering Milestones

first committed to the PAP. This commitment to the PAP must then be carried on down through the organization in order for any impact on the product to be effective.

Product assurance plans specifically involve documentation of tasks, outputs, and quality and reliability methods, which must occur prior to a specified milestone event in the product development process (see Figure 3.2). The assignments consist of engineering activities necessary to build the product design or develop the manufacturing process. These activities may include: (1) conducting finite element analysis (FEA) studies to investigate mechanical or thermal stress effects on a design, (2) performing engineering development tests to evaluate electromagnetic capability (EMC), etc. The outputs consist of the results from various engineering activities. These results may include: (1) engineering development test analysis and report, (2) statistical evaluation of FEA studies, etc. The quality and reliability methods help to provide support in the execution and evaluation of the various work assignments which, in turn, produce the desired output for supporting the engineering milestone event. Finally, design reviews and cost of quality assessment function as checkpoints to the engineering milestone.

There are three levels to a PAP. First, there is a system level PAP to accommodate the top level design of a particular manufacturer. Second, there is a subsystem level PAP to accommodate a primary module of the system to be manufactured. Third, there is the component level PAP to handle the basic elements that would comprise a subsystem of the product to be manufactured. All PAP's would involve similar quality and reliability activities throughout the product development process.

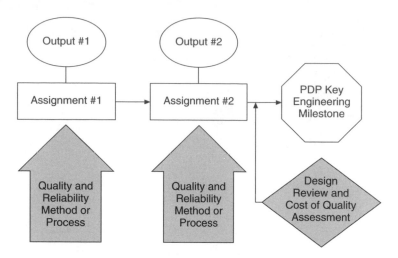

Figure 3.2   Relationship of Product Assurance Method to PAP

The construction of a PAP involves strategic placement of various quality and reliability methods into the sequence of a PDP. Each product development process (PDP) contains a phase with specific program milestone dates, which help to structure major decision targets in the program prior to the next phase in development. The product assurance planning utilizes these principal timing milestones as the end point of each phase in the PDP. Also, the PAP should be kept *short and simple* in order to avoid micromanaging the team (see Figure 3.3 on pp. 28–29). To ensure successful implementation of PAP there should be periodic design reviews of cost of quality management to evaluate: (1) design and process risk, (2) cost of quality allocation, (3) accomplishment of outputs, (4) identification of outstanding issues, and (5) resolution of corrective actions. The design review should be performed by the team members throughout the product development process in order to maintain consistency and focus on the product through its development cycle. This team should be a cross-functional group consisting of executives, managers, product/process personnel, and suppliers. As the product continues through its development, the team may add new members as it moves from a design to a process or assembly activity.

C. Integration of Product Assurance Tools and PAP into the PDP

To achieve product assurance effectiveness throughout the product development process it is important that primary product assurance tools be strategically implemented to support key milestones during a product's development. The relationship of the particular product assurance tool within each phase in the development cycle helps to "build-in" the quality and reliability of the product. It is not so much the use of a tool at any point in the development cycle that is important, rather it is the type, location, and relationship of a PA tool to other design and process engineering techniques that are most effective to improving the product. The following text will discuss the types of product assurance tools and their interrelationship with one another within each phase of the product development process.

Planning and Concept Development Phase Activities

During the planning and concept development phase the concept and program approval milestones are achieved through a series of activities. First, customer expectations are obtained for a desired product. A marketing study is performed to help recognize these expectations. Once these expectations are understood, they are then translated to performance requirements or functional and reliability requirements. After these requirements are determined, key product and process characteristics, which affect the performance of the design, are determined. Second, the product design and manufacturing process evolves through these detailed requirements and characteristics. Finally, from the performance criteria and initial diagrams of the product and process, a test plan is established to address test design and sample size planning for engineering development, design, and process validation.

To effect the transformation of the customer requirements to a product design and process, the use of such PA tools as quality function deployment (QFD) help to logically translate the "what" of a customer expectation to a "how" or product performance criteria. The information obtained about competitor's products such as design and process features from a benchmarking activity should also be used during the QFD exercise to help the team better position the design performance criteria to the competition.

Test sample planning and test design techniques are used to design the test standard, which should accommodate the test sequences and sample sizes necessary for engineering development and verification. These particular PA tools help to direct the product planning toward product performance criteria necessary for the design and process development stage.

Design Development Phase Activities

During the design and development phase program, system and pilot milestones are achieved through a series of activities. While these activities are being performed for the design phase, there are activities being pursued in parallel for the process development phase. The first step of design development is to evaluate the design without testing by using various engineering tools such as simulation (FEA—mechanical, structural, thermal, electrical, fluid, etc.). Second, risk evaluations are performed to address any design flaws from the original design concept. Next, engineering development tests or experiments are performed to determine performance feasibility and allow for any adjustments or redesigns based on actual environmental stresses. Finally, key design performance criteria are addressed to finalize the design prior to design validation testing.

To impact the design development effectively, various PA tools are used throughout the phase. First, a fault tree analysis technique is used to address top event conditions and determine the critical fault path leading to significant risk. In addition, the DFMEA is used to assess, from the bottom up, the components responsible for affecting the subsystem or system. These analyses may be revisited frequently throughout the design development phase in order to include results from simulation or tests that may introduce new risks or reduce potential risks to the design.

Second, a design for manufacturability and assembly is conducted to help review any concerns that may affect the design from a manufacturing or assembly standpoint. These concerns should be verified before engineering development tests in order to gain some initial exposure under actual environmental stresses or processing conditions if tools are available.

Third, quality and reliability methods should be applied during the simulation stage of design development in order to incorporate statistical analysis and confidence to the design. Quality techniques used, such as variation simulation analysis, help to address dimensional tolerancing issues and identify significant input factors that affect product fit, form, and functionality. Reliability techniques should be applied to address reliability performance during engineering development testing. These techniques range from the use of stress/strength evaluation to the use of the Weibull distribution to help determine product life parameters and reliability. Results from each of these studies should be fed back to the DFMEA or FTA in order to reassess risk during the design development process.

Process Development Phase Activities

Similar to design development phase activities there are various process development activities, which are performed to achieve key process milestones. The first step in process development is to establish the process definition and generate a process flow diagram. Based on this diagram an FMEA is performed on the complete process through the evaluation of individual machine operations in order to identify failure modes, current controls, and risks resulting from each process

Product Design: Electronic Module Type C
Manufacturer: Company B (First-Tier Supplier)

Product Development Phase	Description of Product Development Assignment/Task	Completion Dates		Responsibility	Design Review and Cost of Quality Assessment	Engineering Milestone
		Planned	Actual			
Planning and Concept Development	Conduct Marketing Study	1/12/XX	1/8/XX	Company A	Reviews held: 3/24 and 4/14/XX	Concept and Program Approval 4/22/XX
	Perform QFD Study to Identify Customer Requirements	2/15/XX	2/21/XX	Company A, B	Issues:	
	Conduct Benchmarking Analysis	3/22/XX	3/21/XX	Company A, B	Testing sample	
	Establish Competitive Position	4/10/XX	4/20/XX	Company A	size, unit cost,	
	Develop Test Sample Plan	3/15/XX	3/15/XX	Company A	marketing segment (see issue list 4/15)	
Design Development	Generate Functional Block Diagram	3/15/XX	3/13/XX	Company A, B	Reviews Held: 5/12	Prototype 9/10/XX
	Perform Design FMEA and Fault Tree Analysis (FTA)	4/20/XX	4/19/XX	Company B		
	Conduct Design for Manufacturability and Assembly (DFMA) Study	5/2/XX	5/11/XX	Company A, B, C	Issues:	
	Perform Engineering Simulation Studies:				Address (7) major risks identified	
	1) Finite Element Analysis (FEA)	5/23/XX	5/29/XX	Company B	in FMEA and FTA,	
	2) Variation Simulation Analysis (VSA)	6/17/XX			DFMA proposal #3 action plan needed	
	3) Mechanical and Electrical Functional Evaluation	6/20/XX	6/10/XX	Company B		
	Reliability Assessment and Prediction	6/10/XX	6/17/XX	Company B	Total cost of quality	
	Reliability Growth Test Management	6/20/XX	7/15/XX	Company B	= $X (Prevention)	
	Conduct Development Test	10/30/XX	11/29/XX	Company A, B		

Phase	Task			
Process Development	Generate Process Flow Diagram	4/20/XX	4/4/XX	Company B
	Perform Process FMEA	5/5/XX	4/29/XX	Company B
	Develop Process Control Plan	5/29/XX	6/10/XX	Company B
	Determine Measurement System Calibration and GRR	7/2/XX	6/24/XX	Company B
	Perform Tool and Equipment Studies	8/10/XX	8/2/XX	Company B
	Establish Preventive Maintenance Plan	8/15/XX		
	Review Packaging and Shipping Designs	5/15/XX		
	Conduct Process Performance Studies	11/5/XX		
	Operator and Machine Instructions	9/14/XX		
	Conduct Process Review #1	10/23/XX		
	Facility Review	12/5/XX		Pilot 4/10/XY
Design and Process Validation	Perform Design Validation Test	4/3/XY		
	Review DV Test Data Analysis	4/20/XY		
	Perform Process Validation Test	7/10/XY		
	Analyze Process Performance Studies	7/15/XY		
	Review PV Test Data Analysis	7/20/XY		
	Execute Design of Experiments	Ongoing		
	Conduct Process Review #2 (if nec.)	5/2/XY		Launch 12/1/XZ
Production and Continuous Improvement	Perform Statistical Process Control Studies and Analysis	1/25/XZ		
	Conduct Root Cause Evaluations	Ongoing		
	Perform Field Return Analysis	Ongoing		
	Total Cost of Quality Review	2/15/XZ		

Figure 3.3 Product Assurance Plan Worksheet—Example

step. Next, a process control plan is developed that accommodates the current controls identified in the risk analysis and helps establish the details on specific process controls, measurement processes, and corrective action procedures necessary to prevent process abnormalities. From the process control plan, preventive maintenance plans are reviewed for each piece of equipment or tooling involved in the process in order to support the control of the process. Before the preventive maintenance plan is finalized, tool and equipment studies should be performed to understand the life expectation of the tool and the degree of maintainability necessary to prevent equipment failure or process variability. The key process characteristics determined and developed in the control plan are translated into operator instruction sheets and equipment set-up instructions.

The next step in process development includes review of packaging and shipping. These particular activities involve the evaluation of the package design relative to the handling involved during transportation to final assembly. A validation of the package design under actual shipping conditions would be performed just as the product design would be tested during the design validation testing activity.

Once the control plan is detailed, measurement systems are evaluated prior to the execution of initial process performance studies. These preliminary process performance studies are performed on each piece of equipment or tool to establish baseline process potential and capability and allow adjustment of the equipment to assure better capability under production conditions.

The final activities in process development involve a process review. The first process review (process review #1) is done during process development, while the other review (process review #2) is done (if necessary) after process validation testing. Conducting a second review may be done if the process has changed in any degree after testing. A process review should be performed to carefully evaluate each of the previously described process development events and addresses such areas as incoming material qualification, measurement and process performance, parts handling, packaging and shipping, operator instruction sheets, and the identification of the proper process control technique (error and mistake-proofing, variable and attribute control charts, etc.). The intent in a process review is: (1) to assure that the elements of QS-9000 are properly in place and being used effectively in the organization, and (2) to obtain pre-production part approval (PPAP). "The purpose of pre-production part approval is to determine if all customer engineering design record and specification requirements are properly understood by the supplier and that the process has the potential to produce product meeting these requirements during an actual production run."[1]

To impact the process development phase effectively it is important to use the appropriate PA tool. First, the risk assessment is performed using a PFMEA similar to the design PFMEA. This helps determine the priority risks and establishes initial current controls. It is important that the controls identified in the PFMEA become further developed and supported by scheduled equipment preventive maintenance action based on previous tool and equipment life expectancy studies or prior process capability studies. Next, a process control plan should be developed to describe how each product or process characteristic is evaluated.

Based on the identified measurement technique for appraising the product or process characteristic, a calibration and gauge repeatability and reproductibility evaluation is performed to understand the degree of measurement capability. An appropriate measurement system analysis technique should be applied for evaluating either variable or attribute type data. Once the measurement systems are evaluated and gauge error variation eliminated or controlled, process performance studies should be done on each key product and process characteristic per machine/operation in order to understand the relationship of key design/process characteristic variation to the tolerance limits. The use of statistical process indices such as P_p, P_{pk}, $P_p - P_{pk}$ (capability indices for stable processes: C_p, C_{pk}, $C_p - C_{pk}$), or the probability of success should be considered for analysis.

Product and Process Validation

Activities performed in the validation phase involve design and process verification tests. Design verification testing is a validation that a design functions as expected under test. Process verification testing involves validation of fit, form, and function of samples built off of production tools and processes. During each of these tests various quality and reliability tools are applied to help evaluate the results. All fixed environmental stress tests are evaluated using test data analysis of statistical differences of before and after test. Any extended tests are evaluated using either stress/strength or Weibull distribution techniques for life expectancy. The goal is to determine the Weibull parameters necessary for reliability calculation and comparison to reliability targets. Any failures or significant differences in key functional characteristics undergo failure analysis for root cause and subsequent corrective action. This activity may involve further knowledge of materials, electronics, and mechanics for the product assurance specialist to use in suggesting further experimentation such as DOE, etc. Other activities in this phase involve continual updating and review of DFMEA and PFMEA to assess risk and current controls and a process facility review if significant changes were made to the process that would affect process performance.

Finally, process validation test is done to verify product performance based on production intent tooling and assembly. This process validation test should be evaluated using reliability analysis tools such as Weibull life analysis and test data analysis of before and after data. It is important to note the time to failure as well as any significant changes in functional parameters of parts under environmental stress tests. Last of all, a process review should be conducted to help validate that the production intent process is in place and that the quality system which has been established is working effectively to address the specific elements of QS-9000. As mentioned before, if changes are made to the process or design, then it is absolutely necessary to reevaluate the process—see PPAP submission requirements (page 2 of PPAP manual).

Production and Continuous Improvement Phase Activities

The last phase in the product development process involves initiating the launch of the product and the manner of continuous improvement. Basically, the product goes through production at normal line rates with all process controls in place and strict process monitoring. PA processes used involve examination of process control chart behavior and capability studies of the individual operations and total process. Any failures in the manufacturing facility or in the field require careful examination through Pareto charts and then through various analysis techniques such as root cause analysis, or further experimentation such as DOE if the problem is attributed to a design or process feature. All common or special causes of variation are reflected back to the DFMEA or PFMEA for risk reevaluation. The FMEA is the central risk assessment tool that helps to document concerns and developmental activities in place to reduce the current risk. Any corrective actions made in the design or process are revalidated through test and process capability before entering production. A process sign-off may be necessary to obtain PPAP (production part approval process) prior to production.

A total cost of quality assessment is finally performed at this phase to determine the quality cost allocation as well as the total cost of quality. The goal in quality cost management is to have most of the total cost allocated to prevention activities, which would subsequently lower the potential for internal and external failures and decrease the total life cycle cost.

NOTES

1. Chrysler, Ford, and General Motors Supplier Quality Requirements Task Force, *Production Part Approval Process (PPAP)*, 1993, p. 2.

4

Product Assurance Tools and Processes and Their Application
Concept Development Phase Tools

"Man is a tool-using animal, without tools he is nothing, with tools he is all."

THOMAS CARLYLE,
Sartor Resartus,
1833, BK I, CH. 5

Chapter Introduction

This particular chapter is the first of four chapters involving the strategic use of various quality and reliability tools during each of the product development phases. The purpose of this chapter is to discuss the practical application of three product assurance tools and processes used during the concept development phase (refer to the PAP worksheet example, Figure 3.3). These tools involve: (1) design/process requirements through the use of quality function deployment (QFD), (2) benchmarking, and (3) test plan development and sample planning (hardware and software). Strategic implementation of these quality and reliability tools helps to support the product development phase concept approval milestone.

A. Requirement Definition and the Use of Quality Function Deployment (QFD)

What and Why the Tool Is Applied

The key objective of quality function deployment (QFD) is to systematically translate consumer expectations obtained from marketing studies into functional requirements for the product or service. The QFD activity should consider product requirements at each stage from planning and concept development and design/process development to validation, production, and finally marketing/sales and distribution. This method establishes the foundation of product performance requirements. In a way, it is the act of taking the *voice of the customer* all the way through product development, to the factory floor, and then out into the marketplace. Specifically, QFD involves transforming, through a matrix, customer wants (what's) to company measures (how's) and then further transforming them to design, manufacturing process, and production control requirements (see Figure 4.1). The QFD process is well suited to the *simultaneous engineering* concept in which product design and process engineers participate as a team effort. QFD may be thought of as a blueprint for the operation of such product development teams.

When the Tool Is Applied

Quality function deployment is primarily applied during the planning and concept development stage of a product development process. This tool is then applied successively throughout the concept development phase to translate customer requirements to design and process requirements. The development of the design and process requirements is a parallel activity. Further, the transformation of customer requirements will help develop other requirements such as part characteristics and process requirements. This transformation of customer requirements will help develop product standards necessary for the product development and validation phase.

Where the Tool Is Applied

QFD is applied to any design or process complexity level (component, subsystem, or system level) and part status (new, modified, or carryover).

Who Is Responsible for the Method

The product and process engineer, and product planning specialist are primary members of the team who are responsible for the development of the product and process requirements. Product assurance helps engineering and product planning facilitate and consult on the use of the QFD process.

How the Tool Is Implemented

Quality function deployment is performed through the development of a *house of quality* matrix. This matrix helps to guide the flow of information from the "what" to the "how" through the relationship matrix, and then to "how much." Each QFD matrix facilitates the development of subsequent matrices starting from the design requirement matrix to the manufacturing requirement matrix. The "how" of a previous matrix becomes the "what" of the subsequent matrix (see Figure 4.2).

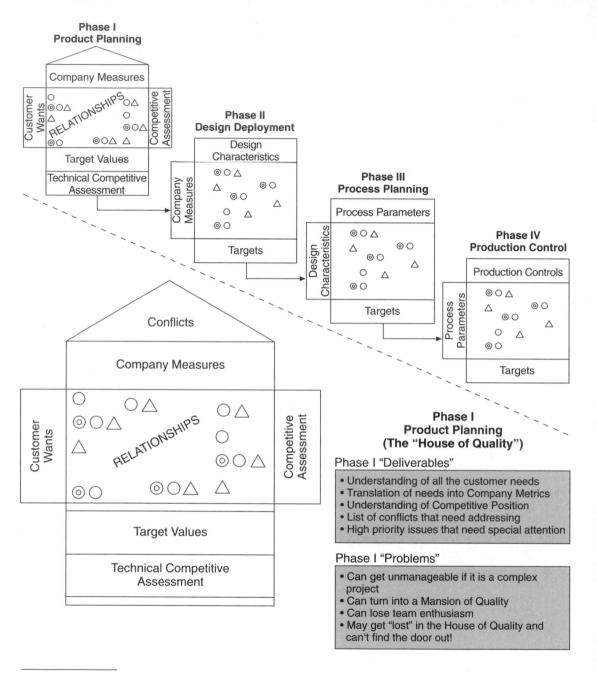

Phase I
Product Planning

Company Measures

Customer Wants

RELATIONSHIPS

Competitive Assessment

Target Values

Technical Competitive Assessment

Phase II
Design Deployment

Design Characteristics

Company Measures

Targets

Phase III
Process Planning

Process Parameters

Design Characteristics

Targets

Phase IV
Production Control

Production Controls

Process Parameters

Targets

Conflicts

Company Measures

Customer Wants

RELATIONSHIPS

Competitive Assessment

Target Values

Technical Competitive Assessment

Phase I
Product Planning
(The "House of Quality")

Phase I "Deliverables"

- Understanding of all the customer needs
- Translation of needs into Company Metrics
- Understanding of Competitive Position
- List of conflicts that need addressing
- High priority issues that need special attention

Phase I "Problems"

- Can get unmanageable if it is a complex project
- Can turn into a Mansion of Quality
- Can lose team enthusiasm
- May get "lost" in the House of Quality and can't find the door out!

Source: © Copyright American Supplier Institute, Inc., *Quality Function Deployment (QFD),* Allen Park, MI. 1996. Reproduced by permission under License No. 970201.

Figure 4.1 Traditional Four QFD Phases

The following highlights the procedure for producing a QFD chart (see example in Figures 4.3a and 4.3b):

1. Develop the customer requirements from input through representatives of marketing, product planning, etc. Enter these characteristics in the "what" section of the QFD chart.

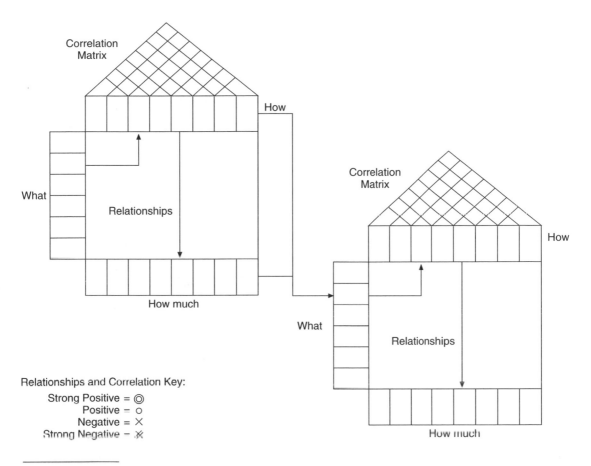

Relationships and Correlation Key:
Strong Positive = ◎
Positive = ○
Negative = ✕
Strong Negative = ✖

Figure 4.2   Quality Function Deployment Technique ("House of Quality")

2. Translate the customer requirements or "what's" into the "how's" or design requirements. Determine the relationship between the "what's" and the "how's" by identifying the magnitude of the relationship.
3. Establish the relationship between the "how's" by determining the degree of correlation and enter in the correlation matrix of the QFD chart.
4. Translate the "how's" to "how much" through engineering analysis. The development of the "how much's" is: (1) to provide an objective means of assuring that requirements have been met, and (2) to provide targets for further detailed development. The "how much's" therefore provide specific objectives, which guide the subsequent design and afford a means of objectively assessing progress. In addition, the "how much's" should be measurable items in order to provide more opportunity for analysis and optimization.
5. Conduct a competitive assessment based on the "what's" and the "how's." The competitive assessment of the "what's" is called a customer competitive assessment and should utilize customer oriented information as much as possible. The competitive assessment of the "how's" is often called a technical competitive assessment and should utilize the best engineering talent to compare competitive products.
6. Apply an importance rating for each what or how to help prioritize efforts and make trade-off decisions. Look upon the number as further opportunities to check current thinking

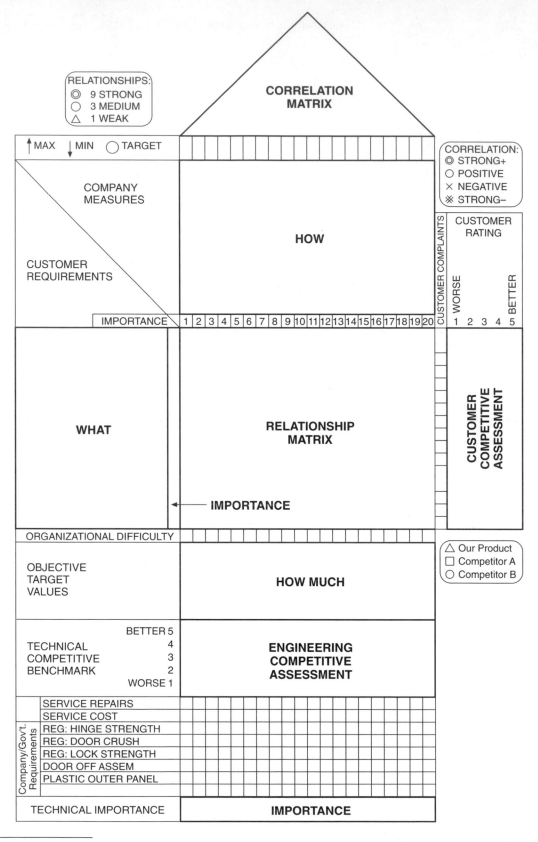

Figure 4.3a   Quality Function Deployment Example

DOOR SYSTEM QFD
PRODUCT PLANNING MATRIX (PHASE1)

COMPANY MEASURES (HOWs) / Column headers:

Door System Measurables:
1. Door Closing Effort O/S
2. Door Opening Effort O/S
3. Door Opening Effort I/S
4. Reach Dist. to Opening Mech.
5. Pull Force I/S
6. Dynamic Hold Open force
7. Static Hold Open Force

Window System Measurables:
8. Reach Dist. to Wdo Open Mech.
9. Grip force needed to crank
10. Hand Clearance
11. Manual Wdo Operating Effort
12. Window Motor Current
13. Amount Water Removed
14. Window Cycling Time (Elec.)
15. Water Leak Amount

Lock System Measurables:
16. Lock/Unlock Time (Outside)
17. Unlock Force
18. Key Insert & Rotation Effort
19. Freeze Resistance

CUSTOMER RATING: △ Our Car   ○ Car "A"   □ Car "B"   (scale 1–5)
Customer Complaints

CUSTOMER REQUIREMENTS (WHATs):

Row	Customer Requirement	Importance (Scale 1-5)	Customer Complaints
1	Easy Close from Outside	5	42
2	Easy Open from Outside	4	
3	Easy Open Inside	3	
4	Easy Close from Inside	4	12
5	Stays Open in Check Position	3	5
6	Handle Looks Nice	2	
7	Crank is Easy to Reach	4	1
8	Crank is Easy to Grasp/Hold	3	14
9	Easy to Operate (Man.)	3	
10	Wipes Dry	2	
11	Operates Rapidly (Elec.)	2	
12	Doesn't Leak Water	5	
13	Lock Knobs Oper. Easily	4	
14	Latch Lasts Long Time	4	14
15	Key Operates Easily	3	
16	Doesn't Freeze	4	36

(Customer requirement groupings: Easy to Open and Close; Window Operates Easily; Lock and Latch Easily — under "Good Operation and Use")

DIRECTION OF IMPROVEMENT (per column)

ORGANIZATIONAL DIFFICULTY (columns 1–19):
4, 2, 2, 4, 3, 2, 1, 3, 2, 4, 2, 1, 3, 4, 3, 2, 2, 2, 5

TARGET VALUES (columns 1–19):
1. 7.5 ft. pounds
2. 15 ft. pounds
3. 8 ft. pounds
4. 26 inches
5. 12 ft. pounds
6. 15 ft. pounds
7. 10 pounds
8. 22 inches
9. 5% female hand-2lbs.
10. 95% male hand
11. 20 inch pounds
12. 15 amps
13. 100 percent
14. 2 seconds
15. 4 hr. spray - 0 leak
16. 0.5 seconds
17. 0.75 pounds
18. 3 pounds
19. 71 hrs. @ -40° F

ENGINEERING COMPETITIVE ASSESSMENT: △ Our Car (5) ○ Car "A" (4) □ Car "B" (3) (scale 1–5)

Frequency of Service: 48 56, 42, 30 3, 30 5, 68 15, 28 17, 34 14
Average Cost of Service

IMPORTANT CONTROLS:
- Reg: Hinge Strength
- Reg: Door Crush Strength
- Door off Assembly
- Plastic Outer Panel

ABSOLUTE IMPORTANCE (columns 1–19):
61, 45, 27, 27, 41, 18, 36, 60, 48, 48, 27, 2, 23, 18, 45, 40, 45, 36, 47

Check which items to deploy to Phase II

Figure 4.3b Quality Function Deployment Example

and make corrections. The end result of the characteristics defined from the "how's" will lead to the development of the "what's" in a subsequent matrix on part characteristics.

7. Construct other QFD charts to address part characteristics, manufacturing process requirements, and production requirements.

Benefits of the Tool

1. QFD provides an effective means of transforming general product/service expectations to specific performance requirements. The method provides a very organized approach to the assembly of product/service information for development.

2. QFD involves the communication and teamwork of many different people from various disciplines in the company organization. This interrelationship of people from product engineering, design, manufacturing, quality and reliability, purchasing, product planning, service, and program management provides the mechanism for simultaneous engineering and input to the QFD process.

3. QFD helps generate a comprehensive list of design requirements for product planning, part characteristics for part deployment, manufacturing operations for process planning, and production requirements for production planning.

4. QFD adapts to the product development process timing and product assurance plan. It is an effective planning tool that has great potential for coordinating and consolidating major planning issues.

5. QFD contains valuable information regarding competitive assessment of specific product design and process requirements as well as the magnitude (how much) of these requirements.

6. QFD establishes priority of engineering development to the obtainment of specific product design and process requirements.

7. QFD facilitates the simplified development of detailed product design and process specifications. This would include performance, design, process, and test standards.

8. QFD helps to organize product design and process information into a main summary table (*house of quality*), which contains several smaller tables of information regarding the relationship and technical importance of the specific design and process elements.

B. Benchmarking

What and Why the Tool Is Applied

Benchmarking involves a dedicated effort to measure your company's processes by those of the best competitors in your industry and outside the industry. The benchmarking activity allows you to continuously learn how others do the process, adapt what you learn to your own company, and initiate action to meet or exceed the best competitor. Benchmarking determines the effectiveness of a competitor by measuring results against reference points. The process of benchmarking forces management to look outside its own operations and outside its organizations to find those companies that are superior in their design and process. With this particular information, performance goals for attaining a leadership position in the business and the implementation of action plans to achieve that position are possible. Some of the goals for benchmarking involve market share, quality and reliability, cost, service, development time, and customer satisfaction. Benchmarking helps the product assurance organization to facilitate the evaluation of the design or process by providing a basis for comparison.

When the Tool Is Applied

Benchmarking is primarily performed throughout planning and concept development and in the early stages of product design and process development. Benchmarking is often used as a check against the proposed design or process during the design and process validation stages.

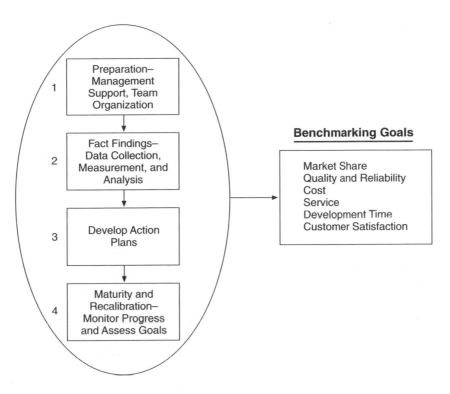

Figure 4.4   Benchmarking Process Steps

Information gained during the early stages of the benchmarking exercise, such as in the planning and concept phase, are applied during product definition (QFD) to facilitate prioritization of design and process development.

Where the Tool Is Applied

Benchmarking is applied to any design or process complexity level (component, subsystem, or system level) and part status (new, modified, or carryover).

Who Is Responsible for the Method

The individuals in the organization who are responsible for implementing the benchmarking process involve marketing, product planning, and product engineering. It is necessary to collect the design and process information, including measurement data regarding functional performance of the product. This may even include testing the product in order to obtain parametric data and reliability information in addition to available warranty and field reports. Management support is critical for benchmarking to be a success. It is ideal if upper management commissions and manages the benchmarking activity, because it knows where the organization is in need of improvement and can guide the benchmarking efforts on reaching those improvement goals.

How the Tool Is Implemented

The benchmarking process involves execution of four basic steps (see Figure 4.4). These steps help guide the evaluation team to efficiently plan the activity, determine the parameters of the design or process to be evaluated, collect and analyze the data, evaluate, and compare the results to the product under evaluation. See the example benchmarking study in Figure 4.5.

A company has committed to perform a benchmarking study on tennis racket designs and processes in order to better guide their product design and process development activity. The following is a brief summary of the study and the comparisons made between competitors.

1. Preparing to Benchmark

   A. Understand the product design/process operation of the present company.

      An intensive study was conducted to review the product design and process development process. The tennis racket design process, including engineering studies (frame material composition and integrity evaluation), simulation (FEA of structural capability), and test planning and analysis (life test evaluation under environmental stress conditions) was reviewed. The manufacturing process development activity was reviewed on the basis of equipment and tool capability, measurement assessment, preventive maintenance, process control, etc.

   B. Review the organizational structure and communication process.

      A review was made to understand the product development process, and the communication channels between engineering functions (design and manufacturing) and company disciplines such as Senior Management, Purchasing, Marketing, Finance, etc. The simultaneous engineering concept was observed and evaluated between the various departments involved in the tennis racket business unit.

   C. Obtain management support.

      The tennis racket engineering team had reviewed with management the need to benchmark their new tennis racket design and process since it is radically different from present designs. Engineering had explained to management the need for design improvement such as: (1) new frame material for greater frame flexure, which would contribute greater ball momentum upon contact, (2) extended "sweet spot" area, and (3) greater handle grip capability. Process improvements included: (1) new frame welding machine and (2) string tensioning calibration.

   D. Devote the time.

      Engineering and management agreed to devote one month of product development time to gather the data and evaluate the results prior to further design/process development of the proposed tennis racket design.

2. Fact Finding

   A. Determine what is to be benchmarked.

Continued.

Figure 4.5 Benchmarking Example

Specific parameters to be evaluated on the new tennis racket design include:

1. Material type and composition
2. "Sweet Spot" location relative to frame opening
3. Handle grip retention and feel

Specific elements to be evaluated on the new tennis racket process include:

1. Frame welding operation—temperature, energy, position, pattern
2. String tensioning operation—equipment calibration and GRR, fixture design, string tension sequencing

Business parameters regarding price and promotion include:

1. Piece cost, indirect and direct material cost, labor, sales price
2. Advertising strategy—medium, location, etc.

B. Determine who is to be benchmarked.

The companies to be benchmarked include Company A, Company B, and Company C.

C. Define the data collection method.

The process of collecting the data will involve examining the strengths and weaknesses of the design and process as compared to product under examination. The mechanisms behind the data gathering include:

1. Teardown analysis
2. Test evaluation under various environmental conditions
3. Material properties evaluation
4. Process flow diagram
5. Process control plan and statistical process control behavior
6. Frame weld machine performance statistics and process capability
7. Product costing information
8. Promotional techniques—location and advertising media—commercials and periodicals

D. Analyze the results.

The analysis of the data collected may take various forms and involve many kinds of analysis techniques. A simple and common analysis methodology involves grouping the data collected according to important product design/process benchmarking factors and assessing the advantage (opportunity or strength) per factor across the various benchmarking candidates. From this matrix approach, specific factors can be compared and an overall assessment can be made as to which candidate has the more favorable design, process, marketing, etc. (see example in Figure 4.6).

Continued.

Figure 4.5 *Continued.*

3. Developing Action Plans

    A. Communicate the results.

       It is extremely important to communicate these results to top management and gain their acceptance before further developmental action is pursued. This benchmarking experience will be wasted unless management is fully committed to support the effort.

    B. Define and implement action plans.

       Based on the results from the study, the design, process, and marketing activities should incorporate these findings into their developmental plans through the requirement definition phase. In the example case, the significant findings per benchmarking factor between companies A, B, and C would be transformed into the QFD study and further transformed into specific design, manufacturing, testing, and service requirements. In particular, the material composition (magnesium/graphite) and mechanical properties (high tensile strength) would be defined in the design specification.

4. Maturity and Recalibration

    A. Monitor and update progress.

       The incorporation of the benchmarking results should be continuously monitored through use of some of the product assurance management tools—PAP, quality and reliability assessment, and the cost of quality. The various design and manufacturing factors used in the benchmarking example would be reexamined in design reviews in order to assess the risk to the customer.

    B. Communicate the results in development to the employees.

       The improvement or redesign factors and the results of the incorporation into the design should be clearly highlighted to the employees so that they can see how the benchmarking process helps to further improve the product and meet or exceed the competition's performance.

Figure 4.5   *Continued.*

The following highlights the four-step procedure for conducting a benchmarking study:[1]

1. Preparing to benchmark first involves *understanding the strengths and weaknesses about the design or process.* The next part in the preparation is to properly document and store the information where it can be quickly retrieved and evaluated. Communication between all parties of the evaluation team is necessary in order to avoid wasted time and effort. It is important that management support exists at the onset of the activity in order to significantly impact the product or process development. The team must be cross-functional and willing to devote the necessary time and effort in collecting information.
2. The *collection of information and analysis* is an extremely important step since it involves the decision as to: (1) what parameters of the design should be evaluated, (2) who is to

Benchmarking Factor	Company A (current company)	Company B	Company C	Comparative Assessment
Frame material composition and strength rating	Aluminum Alloy Moderate tensile strength	Magnesium/Graphite Composite High tensile strength	Laminated Wood Low tensile strength	Advantage B
"Sweet Spot" location and frame coverage area	Top-Center with 55% coverage	Center with 70% coverage	Center with 50% coverage	Advantage B
Handle grip material and handle grip size	Leather/Rubber Composite and 5-inch handle	Rubber synthetic 4½-inch handle	Leather and 4-inch handle	Advantage A
Frame welding process type and process potential/capability	Manual—temp., weld position, energy SPC. 91% FTC	Automatic—robotic controlled 99.7% FTC	Semiautomatic —glue adhesion, 84.5% FTC	Advantage B
String tensioning calibration and measurement capability	20% measurement GRR	5% measurement GRR	32% measurement GRR	Advantage B
Market price range	$80–110	$70–85	$60–80	Advantage C
Promotional techniques	Tennis magazines only	TV commercials, Internet, various sports magazines	Tennis magazines only	Advantage B

Decision—Benchmark to Company B design, manufacturing, and marketing factors.

Figure 4.6   Benchmarking Analysis Matrix Example

be benchmarked, (3) how the data is to be collected and measured, (4) the data collection means, and (5) the analysis of the results. The key element of these steps is the identification of the specific parameter and how it is to be measured. The QFD tool used in product requirement definition helps to identify the product/process characteristic and the importance it has to the rest of the design or process. This tool, which is also used early in the concept development phase, helps to determine the magnitude of the product/process characteristic.

3. The third step in benchmarking involves *developing action plans to utilize the results of the benchmarking study* as discussed in step two and implement any change to the present design or process. The communication of the results to upper management is key to implementing design or process improvement quickly and allowing sufficient time for design and process development.

4. The last step in the benchmarking process is to *monitor the action taken to improve product design and process* to the benchmarking target and update the goal to accommodate new information on the competitor's design and process.

Benefits of the Tool

Benchmarking benefits the organization in many ways. It helps to direct the engineering development activity toward obtaining and exceeding strategic product design/process objectives. Some of these crucial success factors involve management commitment, flexibility, quest for reaching company targets, and trust between those individuals involved in the process.

A summary of the benefits of benchmarking are as follows:

1. Benchmarking promotes initiative to meet customer requirements and address changing customer expectations.
2. Benchmarking encourages the team to strive for success and innovative thinking.
3. Benchmarking allows for a comprehensive understanding of a company's own product design and process as well as its competition.
4. Benchmarking helps to establish realistic goals and measurement parameters for evaluating product design and process.
5. Benchmarking provides an initiative for continuous product/process improvement.
6. Benchmarking helps to achieve benchmarking goals of market share, quality and reliability, cost, service, development time, and customer satisfaction.

C. Test Plan Development, Sample Planning, and Analysis

1. Hardware Model

What and Why the Tool Is Applied

Hardware test plan development and sample planning involves the determination of product development and verification test sequences along with an identification of the number of samples to be used in test for meeting reliability demonstration goals. The test plan and sample requirements are usually contained in a formal document performance test standard, which defines the distribution of parts for test, the test sequence, and the description of test and their objectives. Test plan development is necessary in order to clearly define the sequence of environmental stress types and stress levels to be applied for samples under design/process development and validation. These tests need to be carefully developed to reflect actual field usage, including duty cycle, sample orientation, environmental exposure, etc. Test sample planning involves the statistical determination of the number of samples needed to satisfy reliability demonstration requirements for specific confidence levels. Each test condition during either development or verification testing will require a different analysis method for the determination of the appropriate number of samples.

When the Tool Is Applied

Test plan development and sample planning occurs in the later stages of concept definition and development. It should be developed as soon as the product design is conceived and all requirements are understood. These requirements should include, in addition to the functional design and process requirements, a profile of the component, subsystem, or system environment. This environment would involve the product's interface (mechanical, electrical, chemical, etc.), environmental stress application (temperature, mechanical or electrical load, humidity, vibration, shock, etc.), and duty cycle (some specific percentile usage level for a particular geographic region). Sample planning can take place once the customer has defined the reliability goals to be demonstrated for a specific environmental condition and risk level.

Where the Tool Is Applied

Hardware test plan development and sample planning is applied to any design or process complexity level (component, subsystem, or system level) and part status (new, modified, or carryover). Larger, more complex systems may involve fewer test samples due to test chamber accommodation, expense, etc., however, the same techniques can be applied to analyze the reliability of the samples.

Test Conditions:

* Combined power/temperature cycling—life test portion (see example test profile, Figure 4.8)

* Combined series/parallel test environmental exposures (see example test sequence diagram, Figure 4.9)

* 95th percentile customer usage duty cycle (10-year or 100,000-vehicle-mile duration) for component operation:

* 14,500 component cycles (4000 cycles—mean value, 4700 cycles—Weibull characteristic or 63.2% value)

* 300 hours of component on-time test exposure
  400 hours of component off-time (dormant) test exposure

Reliability Requirements:

* Design failure rate goal = 12.645 failures per million operating hours

* Design reliability goal = 94.35% (exponentially distributed failures assumption)

* Life test confidence level = 70%

* Calculated life test sample size = 22

* Test criteria: Must function electrically and mechanically within the specific functional parameter acceptance criteria (example: 2.5 +/–0.25 Amps of electrical current, 6.2 +/– 0.5 N of force, etc.)—see specification # XYZ.

Figure 4.7   Test Development Conditions and Requirements—Example

Who Is Responsible for the Method

The product engineer and test engineers are the primary individuals responsible for the development of the test plan. The product assurance engineer helps to review and provide input as to the test sequence, duty cycle selection (percentile customer usage), and sample size determination. This test plan would also be reviewed by other engineering personnel to assure that the environments, stress levels, test sample orientation, etc., are properly considered.

How the Tool Is Implemented

Test plan development involves several considerations. The following is a list of key considerations to preparing the test plan:

1. Identify all test environments relative to the product's functional atmosphere. Relate the product's design and process requirements to the selection of the environmental stress type (for example, vibration, mechanical shock, temperature, humidity, etc.).
2. Determine the stress level (for example, temperature, voltage level, mechanical load, etc.) for each environmental type. Consider accelerated test levels which have correlation to extended field exposure.

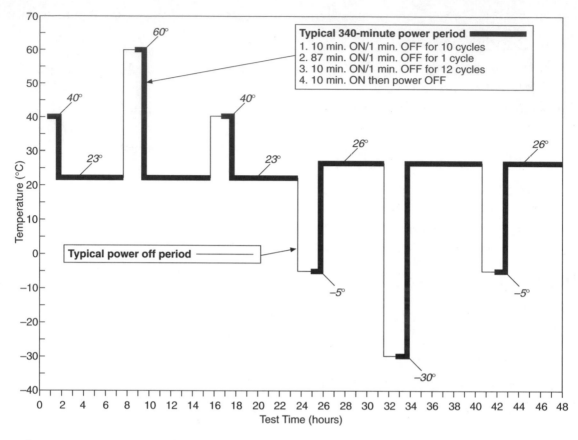

Figure 4.8 Life Test Profile (Combined Power and Temperature)—Example

3. Incorporate the part operating state (power on or off) and duty cycle (for example, number of mechanical switch cycles) into each environmental test situation. Decide on the percentile customer usage level (for example, 50%, 70%, 95%) for the duty cycle and maintain this consistency throughout each test environment stimulus (see examples in Figures 4.7 and 4.8).

4. Randomly arrange the order of the test to consider multiple inputs of the environmental stresses on the same parts throughout the test sequence. Combine tests into one test, if possible, to emulate field usage and environmental interactive effects on the parts (see examples in Figure 4.9). Try to avoid too many parallel tests of individual groups of parts.

5. Determine the length of time or number of cycles for conducting each test. Some test sequences may be structured for reliability verification and others may be extended testing or test to failure.

6. Identify the failure criteria for the test samples. Determine the specification range for functional parameters. Decide on the monitoring frequency for each sample on test. Functional and parametric data may be recorded before and after a test, but may be recorded during a test as well. See Figure 4.10 for an example of a test summary report.

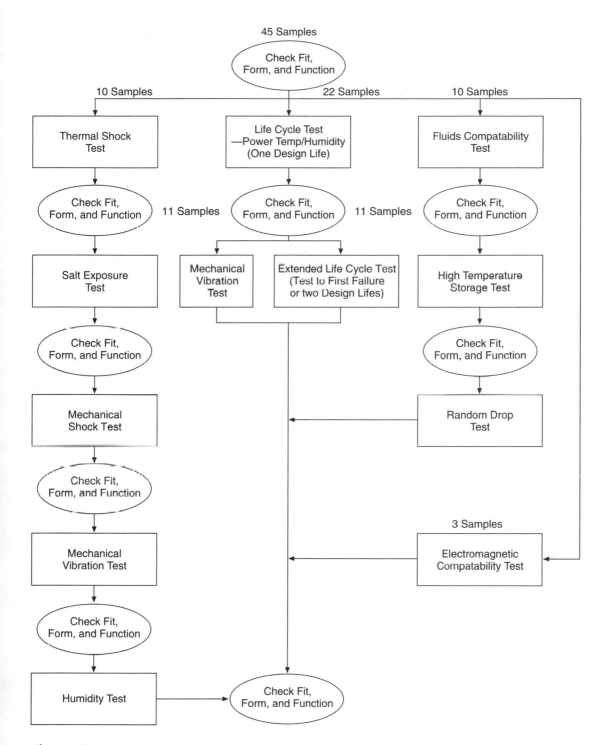

Figure 4.9   Validation Test Sequence (Series/Parallel Arrangement)—Example

Test Type	Performance Specification	Test Description	Pass/Fail Criteria	Test Objective	Test Equipment and Calibration	Number of Test Samples	Results
Thermal Shock	STD-XYZ, Section 5.2	Repid change in temperature: −20°C to 85°C in 30 minutes —operating duration = 20 cycles	—Switch to actuate between 1.5 and 3.0 N of push effort —Electrical voltage drop <200 mV	No failures	Temp. chamber #3, Temp. calibation = ±0.5°C (5/10/87)	10	*Passed*—no statistically significant variation in parameter before or after test
Life Test— one design life	STD-XYZ, Section 5.7	Power on/off— temperature and humidity and vibration application • Temp (−40°C to 85°C • Humidity (95% RH) • Vibration (random schedule— 10 hrs./axis) • operating duration = 30,000 cyc per 400 hours	Same as above	Reliability = 95% for a confidence level = 90%	Environmental chamber #2 • Temp. calibration = ±0.5°C • Humidity calibration = ±2% RH (6/20/87)	45	*Failed*—(R<95%) • 3 failures: 26,500 cyc., 28,850 cyc. and 29,250 cyc. • common failure mode = open circuit • Failure mechanism = worn contact • Actual Reliability = 92.6% (see Weibull analysis)

Figure 4.10 Test Summary Report for Switch Assembly—Example

Sample size determination for development or verification testing is a function of the reliability and confidence level requirement. The following is a summary of the techniques used in determining sample sizes:

Tests used for Reliability Demonstration

1. *Success/Failure Method*

   A1. Success/Failure Method for $r > 0$

   The number of samples is calculated based on the binomial (go–no go status) distribution:[2]

   $$\sum_{x=0}^{r} C_x^n R^{n-x} (1-R)^x = 1-C \qquad (1)$$

   where: n = sample size
   r = number of failures in sample size
   R = reliability (0 – 1.00)
   C = confidence level (0 – 1.00)

   See example #1 of Figure 4.11. A chart for the minimum reliability estimates (R_L) at 90% confidence based on the number of failures versus sample size is shown in Figure 4.12, and a chart for sample sizes (n) at $r = 0$ based on reliability versus confidence level is shown in Figure 4.13.

   A2. Success/Failure Method for $r = 0$

   For special cases when there are no failures ($r = 0$):

   $$R^n = 1 - C \qquad (2)$$

   A binomial distribution may also be used for approximations (see Figure 4.14). An example using the binomial distribution nomograph for $r \geq 0$ can be found in example #2 of Figure 4.11.

   B. Bayesian Method

   A Bayesian approach can also be applied for sample size determination.[3] This method is based on a degree of belief gained by *a priori* probability of a successful product, which has been accumulated through preventive actions and prior knowledge in the design, analysis, and test planning phases of a product assurance program. The confidence gained in a design or process will improve with more application of preventive actions and experience. Since design and process confidence is difficult to measure and quantify, the method tends to be subjective in nature.

   The technique involves identification of the number of samples through an accumulation of the degree of belief and the reliability target and confidence. This degree of belief is accumulated through activities such as reliability plan and analyses, functional and reliability requirements, failure identification and root cause analysis, design and process reliability studies (FEA, benchmarking, stress-strength, life test, etc.), test to field correlation studies, test planning, successful system development, and design/process verification testing, etc.

   A table of sample size based on the prior distribution for the binomial distribution is shown in Tables 4a and 4b. See example #3 of Figure 4.11 for an application of this Bayesian process.

Success/Failure Method Example #1

What is the minimum number of samples with no failures that would be needed to demonstrate a reliability of 95% at a 90% confidence level?

Using equation # 2:

$$R^n = 1 - C$$

we can determine the sample size necessary for demonstration:

$$n = \frac{\ln(1 - C)}{\ln R}$$

$$n = \frac{\ln(1 - 0.90)}{\ln(0.95)} = 45 \; samples$$

Success/Failure Method Example #2

A sample size of 10 parts were tested and 2 failures resulted. What is the reliability for a 95% confidence level? (Assume a binomial distribution.)

Using the binomial distribution nomograph in Figure 4.14 for a sample size of 10, with 2 failures, and a confidence level of 95% results in a demonstrated minimum reliability of approximately 50%.

Bayesian Method Example #3

A system that contains a safety device is under development. The goal is a reliability of 95% at a 90% confidence level. Using Bayesian techniques, what would be a theoretical sample size to test these units?

Phase 1 Activities:	A reliability plan, design, and several reliability analyses were conducted. A test plan was developed, which included a detailed and accurate correlation study of test parameters to field usage. The result of these activities accumulated to a total degree of belief equivalent to 0.25.
Phase 2 Activities:	The system had undergone successful completion of system engineering development testing. The result of this activity accumulated to a total degree of belief equivalent to 0.25.

Continued.

Figure 4.11   Test Sample Planning Examples

Phase 3 Activities: The proposed system had successfully completed design verification testing. All test data was evaluated and any significant variances in the functional parametric were addressed through appropriate reporting and root cause investigation. The result of this activity resulted in a total degree of belief equivalent to 0.20.

Based on the three consecutive successful developmental activities for the proposed design, the sum of the individual developmental phase degrees of belief equated to 0.70 (0.25 + 0.25 + 0.20). This degree of belief approximation corresponds to a medium/high classification after the three successful phases of development activity and would coincide to a sample size of 6 units necessary for production verification testing (see Tables 4a and 4b).

Extended Testing Method Example #4

A proposed motor design is being developed and has a specified reliability target (R) of 95% at a 90% confidence level (C) for a 500-hour operating duration. The engineering team has 8 prototype samples available for life testing. The question is how long should these samples be tested (x_1) without failure to verify the predescribed reliability objectives? This type of motor design does exhibit brush wear at some point in its useful life. Previous Weibull analysis studies that were conducted had shown a brush wear failure mode distribution with a Weibull slope (b) of 3.0.

Using equation # 3, the extended testing duration can be determined:

$$x_1 = x_2 \left(\frac{\ln(1-C)}{(n_1+1)\ln R} \right)^{\frac{1}{b}}$$

Given the above conditions:

$$x_2 = 500 \text{ hours}$$
$$n_1 = 8 \text{ samples}$$
$$R = 0.95$$
$$C = 0.90$$
$$b = 3.0$$

The extended testing duration can be calculated:

$$x_1 = 500 \left(\frac{\ln(1-0.90)}{(8+1)\ln(0.95)} \right)^{\frac{1}{3}} = 854.29 \text{ hours}$$

Figure 4.11   *Continued.*

Sample Size (n)	Minimum Reliability (R_L) Estimate Relative to Number of Failures, r								
	0	1	2	3	4	5	10	15	20
1	10								
3	46	20	3						
5	63	42	25	11	2				
6	68	49	33	20	9	2			
10	79	66	55	45	35	27			
11	81	69	58	49	40	32	1		
12	83	71	61	52	44	36	5		
15	86	76	68	61	54	47	17		
20	89	82	76	70	64	59	34	13	
22	90	83	78	72	67	62	39	19	2
25	91	85	80	75	71	66	45	27	10
50	95	92	90	87	85	82	71	60	50
100	98	96	95	93	92	91	85	79	74
125	98	97	96	95	94	93	88	83	79
150	98	97	96	96	95	94	90	86	82
200	99	98	97	97	96	95	92	90	87

Note: Minimum reliability represents the lower bound estimate (binomial distribution approximation):[2]

$$R_L = \frac{n-r}{\left[(n-r) + (r+1) \times F_{1-C,\ 2(r+1),\ 2(n-r)} \right]}$$

where: n = sample size
            r = number of failures
            C = confidence level
$F_{1-C,\ 2(r+1),\ 2(n-r)}$ = a value with the upper tail equal to $1 - C$ from the F distribution and $2(r+1)$ and $2(n-r)$ degrees of freedom.

Figure 4.12   Minimum Reliability Determination for a 90% Confidence Level from the Number of Failures and Sample Size

TABLE 4a

Bayesian Quantification of Past Experience

Consecutive Successful Test Results	Degree of Belief	
First Success	Low	0.5
Second Success	Medium	0.6
Third Success	Medium High	0.7
Fourth Success	High	0.8

Source: Chrysler Engineering Quality and Reliability, *Test Sample Planning*, Dec. 1990, p.23. Adapted by permission.

TABLE 4b

Reliability Demonstration Sample Size Requirements (Based on Bayesian Techniques)

Reliability Target	Classical Statistics Sample Size (@ CL = 90%)	Sample Size/Degree of Belief			
		Low	Medium	Med./High	High
95%	45	13	9	6	4
90%	22	11	8	5	3
85%	15	9	7	5	3
80%	11	8	6	4	2

Source: Chrysler Engineering Quality and Reliability, *Test Sample Planning*, Dec. 1990, p.24. Adapted by permission.

Confidence Level (C)	Sample Size (n) Relative to Reliability Target, R										
	50%	60%	70%	75%	80%	85%	90%	95%	99.0%	99.5%	99.9%
99%	7	10	13	17	21	29	44	90	459	919	4603
95%	5	6	9	11	14	19	29	59	299	598	2995
90%	4	5	7	9	11	15	22	45	230	460	2302
85%	3	4	6	7	9	12	19	37	189	379	1897
80%	3	4	5	6	8	10	16	32	161	322	1609
75%	2	3	4	5	7	9	14	28	138	277	1386
70%	2	3	4	5	6	8	12	24	120	241	1204
60%	2	2	3	4	5	6	9	18	92	183	916
50%	1	2	2	3	4	5	7	14	69	139	693

Note: Sample size (n) was determined using equation #2 (no failures condition, $r = 0$) of the success/failure method

section A2: $n = \dfrac{\ln (1-C)}{\ln R}$

Figure 4.13   Sample Size Determination from Reliability and Confidence Level (binomial distribution approximation)

2. *Extended Testing Method*

The number of samples determined under this method involves a tradeoff between sample size and test time. Extended testing is a method to reduce sample size by testing the samples to a time that is higher than the test period objective[4]:

$$R^{(n_1+1)(\frac{x_1}{x_2})^b} = 1 - C \tag{3}$$

where: x_1 = extended test time

n_1 = number of samples without failure to x_1

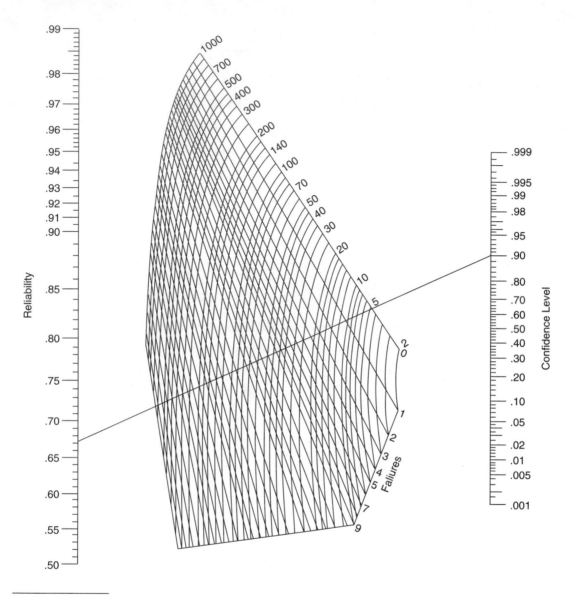

Source: "Larson's Binomial Distribution Nomograph," *Journal of Quality Technology,* Vol. 14, No. 3, July 1982, p. 116. Adapted by permission.

Figure 4.14 Binomial Distribution Nomograph

x_2 = period objective
n_2 = number of samples without failure to x_2
R = reliability
C = confidence level
b = Weibull slope

See example #4 of Figure 4.11 for application of this technique.

3. *Test to Failure*

   The number of samples used under this particular method could be three or more. This method uses ranked time-to-failure data to determine the solution to the Weibull reliability equation. A sample size of three or more would be necessary to achieve a Weibull slope which represents the general behavior for some failure mode distribution. Either rank regression or maximum likelihood estimation methods could be used to obtain the best line fit to the ranked time-to-failure data (see chapter 6, section B, parts 2 and 3 for further details on the use of the Weibull distribution for reliability analysis).

Benefits of the Tool

Hardware test plan development and sample planning is an important activity during concept development. It helps set the stage for the testing program to be conducted throughout the design/process development and validation phases. Some key benefits are:

1. Test plan development guidelines assist in establishing meaningful test sequences prior to the final selection of a performance test standard.
2. Test sample planning helps to consider the minimum number of test samples for meeting reliability requirements. In addition, test sample planning helps to minimize the appraisal cost of quality by initiating careful consideration to efficient testing techniques.
3. The use of test plan development and test sample planning helps to select the best testing alternative for meeting product design and process reliability requirements.
4. These tools and techniques provide a logical and statistical approach to test planning development. They help to avoid generating tests and test plans that may not provide any value or confirmation of potential risks.
5. The execution of test development and sample planning promotes awareness to the testing conditions and the purpose behind the test prior to the busy period of product design and process development.

C. Test Plan Development, Sample Planning, and Analysis

2. Software Model

What and Why the Tool Is Applied

The objective of software test development and analysis is the use of a simple and practical test approach and reliability model for evaluating software test data without including time. This *non–time-dependent model* enables the software engineer to readily apply results from a software test matrix and monitor progress on an ongoing basis during a product development program.[5] Evaluating software reliability test data analysis is an important activity of the overall software assurance process.

Analysis techniques to be presented address software assurance concerns and testing issues by providing (1) a test matrix approach for error detection coverage, (2) a measure of test effectiveness based on the actual test combination executed and the total number of test combinations, (3) a calculation for the probability of success from the test combinations executed, and (4) timely software reliability growth progress. Performing such analysis helps to guide efforts in measuring software conformance to customer requirements.

When the Tool Is Applied

During phase two of the software design and verification stage (see Figure 4.15) a test plan is developed and implemented from the software requirement specification. It is during software test plan development and execution that the reliability model is applied to help assess the magnitude of software reliability and initiate corrective action process leading to higher software reliability.

Where the Tool Is Applied

Software test plan development and analysis is applied to any software revision (new, modified, or carryover).

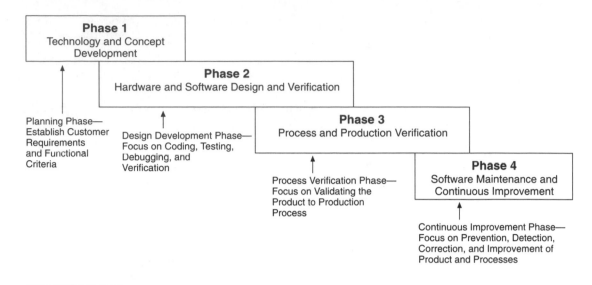

Source: John Bieda, "Software Reliability: A Practitioner's Model," *Reliability Review,* a publication of the Reliability Division, ASQC, Vol. 16, No. 3, June 1996, p. 19. Adapted by permission.

Figure 4.15   Four Phases of Software Development

Who Is Responsible for the Method

The software releasing engineer, test engineer, and assurance specialist work together in this phase to establish the test plan and test matrix to be executed relative to prescribed software input factors.

How the Tool Is Implemented

The strategy for the development and implementation of a software reliability model once software requirements and a test plan are established is shown in Figure 4.16.

1. First, a software test matrix must be determined based on the operating inputs and zone of operation. This matrix can be developed using Taguchi and classical design of experiment (DOE) techniques to help cover ranges in the software test domain. The matrix includes the identification of software input test factors (for example, code for robotic servo motor movement), levels per software test factor (high, medium, and low), and the number of operating zone elements (software input range possibilities) necessary for the selection of a particular test matrix. Each test run number represents a specific treatment combination of factors and levels for which the software is executed and evaluated as to the proper output response.

2. The second step in the software reliability model and execution phase is to determine test matrix effectiveness. This parameter is determined by formulating a ratio of the actual number of software input combinations tested from the DOE matrix and the theoretical total number of input combinations. The test effectiveness ratio helps us to understand the magnitude of test coverage. The higher the ratio the greater the potential of the proposed test to uncover possible software errors.

    Obtaining the test effectiveness ratio involves first calculating the total number of test combinations using the following formula:

$$\text{Total Number of Test Combinations} = cb^a \tag{4}$$

where: a = number of factors
         b = number of factor levels

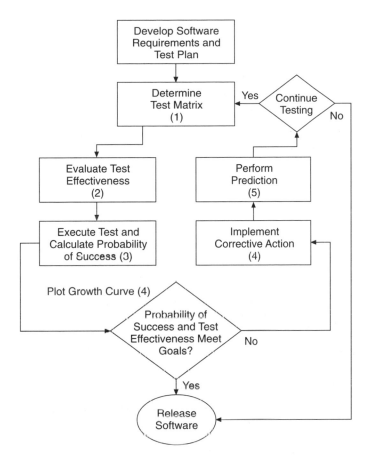

Source: John Bieda, "Software Reliability: A Practitioner's Model," *Reliability Review,* a publication of the Reliability Division, ASQC, Vol. 16, No. 2, June 1996, p. 21. Adapted by permission.

Figure 4.16   Software Reliability Test Development and Analysis Sequence

c = number of operating zone elements (Figure 4.17)
—nominal
—between nominal and upper specification limit (USL)
—between nominal and lower specification limit (LSL)
—above specified limit (out of range)
—below specified limit (out of range)

Using the total number of treatment combinations from the DOE matrix as described in equation 4, test effectiveness is calculated as follows:

$$\text{Test Effectiveness} = \frac{n}{cb^a} \tag{5}$$

where: n = number of treatment combinations from DOE matrix

3. The third step in the software reliability analysis approach is to test the software by executing the DOE test matrix and accumulating output response data for the purpose of determining the probability of success for a specific test design. This probability of success

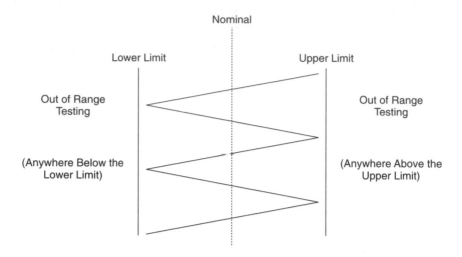

Source: John Bieda, "Software Reliability: A Practitioner's Model," *Reliability Review,* a publication of the Reliability Division, ASQC, Vol. 16, No. 2, June 1996, p. 20. Adapted by permission.

Figure 4.17   Zone of Operation for Software Inputs

is calculated by forming a ratio of the number of acceptable output responses to the total number of treatment combinations:

$$P = \frac{\text{Number of Successful Responses}}{\text{Total Number of DOE Treatment Combinations}} \qquad (6)$$

The probability of success and test effectiveness together produce a measure of the software's ability to successfully produce an output response. During each test phase a software change or new software function requirement can affect the test matrix design. Consequently, the test effectiveness ratio will change due to the revised number of treatment combinations in the test matrix.

4. The fourth step in the software analysis approach is to plot the probability of success along with the test effectiveness ratio and compare these results to customer expectations. The plot represents a growth curve based on the number of software errors detected and corrected per the specific test phase during software development. Use of the growth curve technique provides an effective means of graphically displaying to management the progress of software development and the rate of software error correction. A decision is made at this point as to whether the probability of success and test effectiveness at a particular test iteration meets expectation. If the software meets expectation, then it can be released for production. Otherwise, corrective action needs to be implemented, and the software is then sent through the complete test and analysis loop until it meets the probability of success and test effectiveness goals.

5. A fifth and final step in the software analysis approach is the ability to predict potential software errors in subsequent test phases. A decision is made as to whether to continue testing or release the software based on number of software errors that occurred in the previous test. The prediction of future software errors allows management to monitor the improvement progress relative to the test phase. It must be cautioned that the prediction

technique not be applied too far in advance, since changes in software requirements (new software input functions) and the software test domain may change the overall curvature of the software growth curve. A prediction for subsequent errors can be found using the following equations:

$$\alpha = \frac{\log\left(\dfrac{C_1}{C_0}\right)}{\log\left(\dfrac{M_1}{M_0}\right)}$$

(7)

$$C_1 = C_0 \left(\frac{M_1}{M_0}\right)^{\alpha}$$

(8)

where: α = growth index
C_0 = previous period cumulative errors
C_1 = future period cumulative errors
M_0 = previous period (test #, weeks, hours, etc.)
M_1 = future period (test #, weeks, hours, etc.)

This model is based on several assumptions. First, software errors are independent. Next, corrective actions must have taken place to remove or correct errors, otherwise the growth curve will erroneously reflect improvement. Last, the cumulative period reflects incremental test iterations in software development where new tests are performed to evaluate the software's ability to meet requirements.

The following is an example of this process:

A designed experiment was conducted to evaluate software for an electronic display device based on four factors, three levels, and three operating zone elements. This experiment involved 27 test combinations (L9 matrix performed three times for each operating zone element = 9 × 3 = 27). The operating zone considered was a varying time element between functions, specifically logic reset time. An error was defined as an electronic display outside the prescribed specification limits.

A sample of the L9 (four factors at three levels each) for one operating zone element after test iteration 2 is as follows:

Treatment Combination	Factor A	Factor B	Factor C	Factor D	Error Response
1	1	1	1	1	No
2	1	2	2	2	No
3	1	3	3	3	No
4	2	1	2	3	Yes
5	2	2	3	1	No
6	2	3	1	2	No
7	3	1	3	2	No
8	3	2	1	3	No
9	3	3	2	1	No

The second step in the software reliability analysis approach is to define the total number of possible test combinations. This calculation is illustrated as follows:

$$\text{Total Number of Test Combinations} = cb^a = 3 \times (3)^4 = 243 \qquad (9)$$

where a = number of factors = 4
   b = number of factor levels = 3
   c = number of operating zone elements = 3

Once the total number of test combinations is determined from using equation 9, the test effectiveness ratio can be computed:

$$\text{Test Effectiveness Ratio} = \frac{n}{cb^a} = \frac{27}{243} = 0.11 \qquad (10)$$

This test effectiveness ratio indicates that approximately 11 percent of the total number of test combinations are currently being tested to find possible errors. The use of an automated test system would enable execution of more, if not all, of the total number of test combinations in the software input operating domain.

The third step in conducting the software test analysis was to execute the given matrix and determine the probability of success. Corrections were made to errors discovered after each test in order to improve the software code. The results of four test iterations for the same matrix are displayed as follows:

Test Iteration (M)	Cumulative Errors (C)	Error Rate ($\Delta C/\Delta M$)	Probability of Success (P)	Test Effectiveness Ratio (R)
1	8	8	19/27 = 0.704	0.11
2	9	1	26/27 = 0.963	0.11
3	9	0	27/27 = 1.000	0.11
4	9	0	27/27 = 1.000	0.11

Careful evaluation of the results indicates software reliability improvement throughout each test iteration. As more effective corrective actions were made to the software code, the probability of success increased for the same test matrix. Since no new software functions or changes to the software test matrix were incorporated, the test effectiveness ratio remained the same for each test iteration. However, under actual testing environments test effectiveness could change due to the addition of new requirements or the desire to test more software input combinations.

The fourth step in the software test analysis approach was to plot cumulative errors and the probability of success versus test iteration and determine whether software meets goals. These plots can be found in figures 4.18a and 4.18b. As shown in the graphs, the cumulative errors in Figure 4.18a reached a constant value and the probability of success in Figure 4.18b approached 100 percent as errors were removed and the proper corrective actions implemented. The test effectiveness was equivalent to 0.11 for each test iteration. The primary objective in software reliability growth management is to obtain a test effectiveness ratio equal to 1.00 and a probability of success equivalent to 100 percent for the software under test. In other words, all possible software input conditions would have been tested with no errors. Since the test effectiveness was less than 1.00, a decision would be to further test additional combinations in order to gain more assurance that the software will perform to any combination of inputs.

The fifth and final step in the software test analysis approach was to predict subsequent errors for a future test period. In this example, information on the results of the first and second test iteration

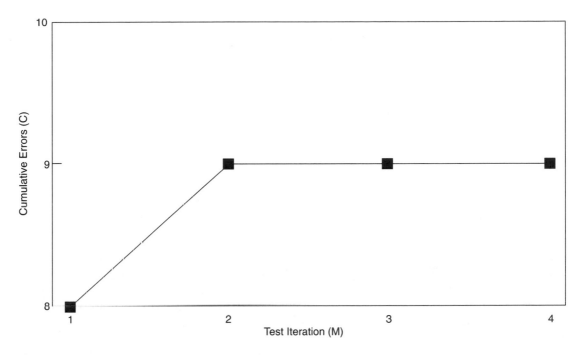

Figure 4.18a   Cumulative Errors vs. Test Iteration—Example

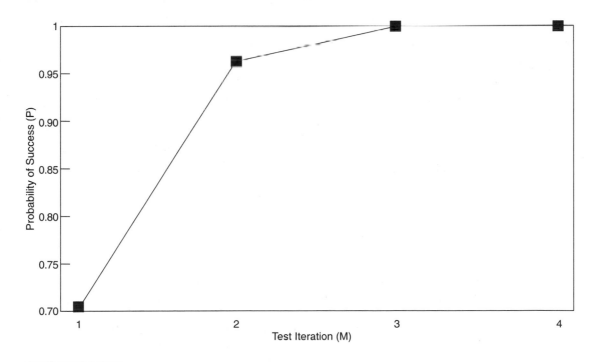

Source: John Bieda, "Software Reliability: A Practitioner's Model," *Reliability Review,* A publication of the Reliability Division, ASQC, Vol. 16, No. 2, June 1996, p. 24. Adapted by permission.

Figure 4.18b   Probability of Success vs. Test Iteration—Example

were used to predict the software error behavior in the third test iteration. From the results, which were evaluated and summarized in the previous table, the growth index α was calculated as follows:

$$\alpha = \frac{\log\left(\dfrac{C_2}{C_1}\right)}{\log\left(\dfrac{M_2}{M_1}\right)} = \frac{\log\left(\dfrac{9}{8}\right)}{\log\left(\dfrac{2}{1}\right)} = 0.17 \tag{11}$$

Use the growth index result from equation 11 to calculate the predicted or future number of cumulative errors in test iteration 3:

$$C_3 = C_2\left(\frac{M_3}{M_2}\right)^{\alpha} = 9\left(\frac{3}{2}\right)^{0.17} = 9.64 \text{ cumulative errors}$$

$$\tag{12}$$

where: α = growth index = 0.17
C_2 = previous period cumulative errors = 9
M_2 = previous period = test iteration 2
M_3 = future period = test iteration 3

Since there cannot be a fraction of an error, the correct result after rounding down would be nine cumulative errors. This predicted result agrees with the cumulative errors displayed in the table for test iteration 3.

Benefits of Technique

1. Software test planning enables the use of a simple non–time-dependent mathematical model for evaluating software test data.
2. Software test planning accounts for a measure of success probability and test matrix effectiveness.
3. Software test planning helps to monitor software reliability as a function of corrective action.
4. Software test planning provides an easy step-by-step method for analyzing software reliability test data.
5. Software test planning enables reliability improvement to be clearly identified and measured.
6. Software test planning utilizes a designed experiment approach for developing the test matrix and considering software input and output response domain.

NOTES

1. The Juran Institute, *Benchmarking,* Quality Management Report, 1994.
2. Lipson, C., and N. J. Sheth, *Statistical Design and Analysis of Engineering Experiments* (New York, NY: McGraw Hill, 1973).
3. Bonis, A. J., "Bayesian Reliability Demonstration Plans," Annals of Reliability and Maintainability Conference, 5th. Conference, 1966.
   Kececioglu, D., *Practical Reliability Demonstration (Bayesian),* Lecture Notes, The University of Arizona, 1985.
4. Lipson, C., and N. J. Sheth, *Statistical Design and Analysis of Engineering Experiments* (New York, NY: McGraw Hill, 1973).
5. Bieda, John, "Software Reliability: A Practitioner's Model," *Reliability Review,* Vol. 16, No. 2 (Cool, CA: Reliability Division of ASQC, 1996), pp. 18–24.

5

Product Assurance Tools and Processes and Their Application

Design/Process Development Phase Tools

"Cease dependence on inspection to achieve quality. Eliminate the need for inspection on a mass basis by building quality into the product in the first place."

DEMING'S THIRD POINT
FRANK PRICE, *Right Every Time,*
ASQ QUALITY PRESS, 1990, P. 53.

Chapter Introduction

This chapter is the second of four chapters involving the strategic use of various quality and reliability tools during each of the product development phases. The purpose of this particular chapter is to discuss the practical application of several product assurance tools used during the product design/ manufacturing process development phases of the product development process (refer to the PAP worksheet example, Figure 3.3). These tools encompass: (1) design and process block diagrams, (2) risk analysis, (3) design for manufacturability and assembly (DFMA), (4) variation simulation analysis (VSA), (5) process control plan development, (6) measurement system analysis, (7) preventive maintenance, (8) reliability growth management, and (9) design review guidelines. The product design and process development phase is a key period in the product development process in which the use of quality and reliability techniques help to detect, analyze, and promote product design and process improvement prior to product validation. Strategic utilization of these tools facilitates the development and approval of the product prior to prototype and pilot engineering milestones.

A. Design and Process Block Diagrams

What and Why the Tool Is Applied

Block diagrams are used to simplify the functional description of either a design or a process. They help to organize the relationship of components or operations to one another in order to illustrate the functional behavior between parts and subsystems. These diagrams provide the necessary description for further engineering analysis tools such as FMEA, FTA, simulation, etc. The design block diagram illustrates the electrical and mechanical functional relationship of parts and subsystems. The process block diagram or "process flowchart" illustrates the relationship between various tools, equipment, and manual operations for the manufacture and assembly of a particular product design. Other block diagrams which involve part or process descriptions may include reliability block diagrams, fault tree networks, etc.

When the Tool Is Applied

Design and process block diagrams are initially generated during the later stages of concept development. Once the product design and process requirements are determined the part, subsystem, or system is illustrated. These drawings are continuously updated during the product design and process development phases as a result of engineering design or process changes.

Where the Tool Is Applied

Design and process block diagrams are constructed for any design complexity level (component, subsystem, and system) and for either new or modified product design and process.

Who Is Responsible for the Method

The product design and process engineer are the primary members of the engineering team responsible for the construction of a design or process block diagram.

How the Tool Is Implemented

Product design and process block diagrams evolve directly from design and process requirements through QFD or other requirement transformation processes. The development of a block diagram involves the following steps:

Functional Block Diagram for a Design

1. Generate a engineering sketch of the proposed design and show its components.
2. Incorporate any specific design requirements into drawing by translating the feature or functional aspect into a engineering sketch. Redraw the proposed design to include these adaptations.
3. Divide the drawing into the components that comprise the proposed design. Illustrate these components through the use of block symbols.
4. Show the functional relationship of each component to adjacent components of the design by illustrating the type of mechanical, electrical, chemical, etc., interface (see example in Figure 5.1).
5. Extend the use of the functional block diagram by transforming it to a reliability block diagram based on a series/parallel failure effect on the system (see example in Figure 5.2).

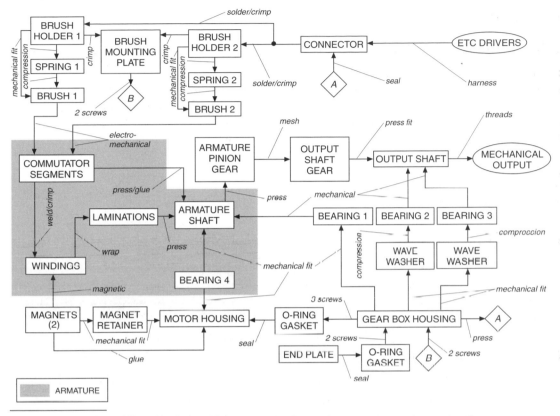

Source: John Bieda and Drew Hoelscher, "Comprehensive Design Reliability for the Automotive Component Industry via the Integration of Standard Reliability Methods" SAE Technical Series, #910357, Feb. 1991, p. 3. Reprinted with permission.

Figure 5.1   Functional Block Diagram—A DC Motor Example

Process Flow Chart

1. Identify the basic process sequence and steps necessary in completing a manufacturing operation.
2. Incorporate any specific process requirements into the process sequence by translating the specific processing considerations. Redraw the proposed process diagram to include these adjustments.
3. Illustrate the process sequence(s) using standard block diagram notation. These symbols should include operation, process control, inventory, shipping, etc., (see Figure 5.3).
4. Show any special transfer means between manufacturing operations.
5. Extend the use of the process flow chart to help evaluate process potential and capability for the total process.

Benefits of the Tool

The following are a list of advantages and applications of design and process diagrams:

1. Provides a basic visualization of the components to a design or process

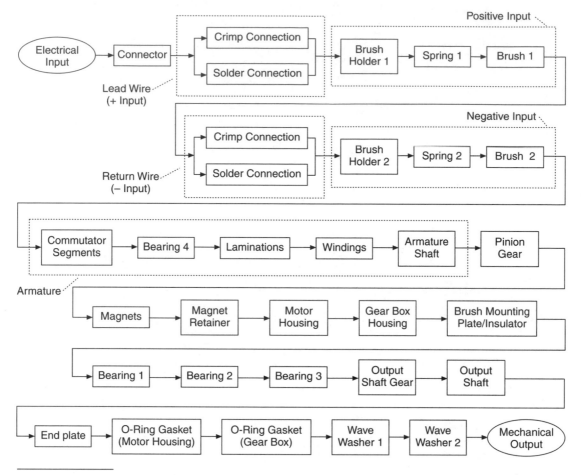

Source: John Bieda and Drew Hoelscher, "Comprehensive Design Reliability for the Automotive Component Industry via the Integration of Standard Reliability Methods," SAE Technical Series, #910357, Feb. 1991, p. 3. Reprinted with permission.

Figure 5.2   Reliability Block Diagram—A DC Motor Example

2. Becomes a reference for various engineering analysis tools such as FMEA's, reliability prediction, simulation, process capability studies, etc.
3. Allows for convenient discussion of product design and process

B. Risk Analysis

1. Failure Mode and Effect Analysis (FMEA)

What and Why the Tool Is Applied

The objective of a Design Failure Mode and Effects Analysis (DFMEA) or Process Failure Mode and Effects Analysis (PFMEA) is to evaluate the magnitude of risk relative to component, subsystem, or system design or process failure modes, and help identify the appropriate control parameter necessary to prevent the occurrence of failure. The analysis is done to help understand the magnitude of risk present in the design or process so that corrective action may be taken to improve on the con-

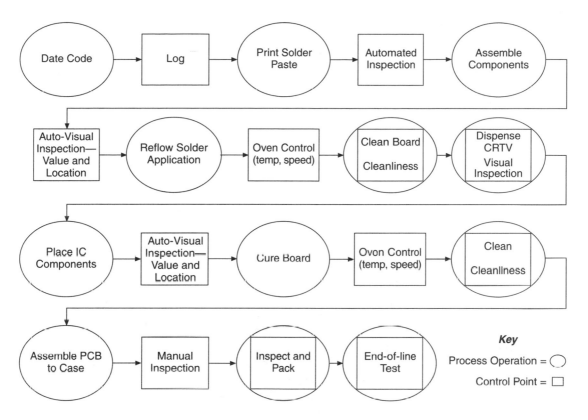

Figure 5.3   Process Flowchart—An Electronic Module Example

trols necessary to prevent failure to the design or process. This type of analysis is a bottom-up form of analysis in that it addresses failures from individual components of a device or operation of a process that may promote risk to the functionality of the subsystem or system. The goal is to obtain no significant risks for any failure mode or failure mechanism. FMEA's are evaluated on a spreadsheet that lists the component, subsystem, or process operation function, failure mode, failure mechanism, effect, current control, risk priority number, corrective action, and responsibility.

When the Tool Is Applied

The DFMEA and PFMEA are developed at the beginning of the design/process development phase and periodically updated throughout the development phase as new information is received from simulation studies, engineering development tests, field and warranty data, manufacturing reports, etc. Both analyses should be updated prior to a design review since this is the session in which issues regarding design and process risk are discussed.

Where the Tool Is Applied

Both the DFMEA and PFMEA are applied to any design or process complexity level (component, subsystem, and system) and part status (new, modified, or carryover). It is important to recognize any minor alteration of the design and process which could introduce a new level of risk to the design or process or create other risks to interfacing components or processes.

Who Is Responsible for the Method

The product design and process engineer are the primary members of the team who evaluate and update the risks as new ones are determined or old ones are altered due to design or process improvements. Product assurance helps engineering review the design and process FMEA relative to other inputs from engineering tests, simulations, and field data, which help validate the level of risks assessed to component, subsystem, or system.

How the Tool Is Implemented

The FMEA for a design or process is performed the same way and uses a spreadsheet as shown in the example (Figure 5.4a and 5.4b). The analysis begins by referring to a functional block diagram or sample of a cut out of the physical part and brainstorming all possibilities of various failure modes and mechanisms of failure for each component, subsystem, or process.

The following highlights the procedure for conducting the analysis[1]:

1. Describe the function of each component, subsystem, or process operation of the product.
2. Determine failure modes of each component or operation. Failure modes refer to the way the device or process fails to perform its intended function.
3. Identify failure mechanisms for each failure mode. Failure mechanisms refer to the root cause of the failure. This points the way to a corrective action necessary for developing a new current control or modifying an existing control.
4. Describe the effect of failure for each failure mechanism. The effect of failure refers to the outcome of the occurrence of the failure mode on the system.
5. Once the failure mode, mechanism, and effect are described, a current control is identified. Design controls are defined as features in the design that affect fit, form, or function. Examples may involve specific parameters from design guidelines, material specifications, dimensional adjustment of mechanical fit, reconfiguration of existing components, variation simulation studies, etc. Process controls are defined as machine adjustment or set-up parameters, which affect fit, form, or function. Examples may be the inclusion of process control check parameters, error and mistake proofing, process simulation studies, gauge repeatability and reproducibility, etc.
6. Calculate the risk priority number based on the product of probability of occurrence, severity, and probability of detection index. The probability of occurrence (likelihood of failure mode occuring) and detection (likelihood of defect reaching the customer) indices are based on a range of low to high probability values. The severity index is correlated to a number which relates to the seriousness of the defect on the system. A range of critical values for the risk priority number would be necessary in order for the analyst to determine when to react and take action.
7. Determine the corrective action needed to develop the appropriate current control for preventing the potential failure mechanism.
8. Identify the responsible individual and date for the corrective action.

Benefits of Technique

1. The FMEA allows for the opportunity to define design and process control features of the product and help prevent potential product design and process risk.
2. The FMEA becomes a central document for the output of many design/process development activities (see Figures 5.5 and 5.6). This document should be the main database of product risk knowledge based on activities performed during engineering development.
3. The FMEA functions as a link from the PFMEA to the process flowchart, control plan, process performance, and capability studies, and process verification testing.
4. The FMEA functions as a link from the DFMEA to the part drawing, development or verification testing, simulation studies, and other design aids.
5. The FMEA promotes teamwork and communication between various engineering disciplines.
6. The FMEA becomes input to the team during design reviews and for risk assessment prior to the key engineering milestone event.

Subsystem Name: Widget
Application Year: 19XX
Supplier Responsibility: RST Engineering

Part Name	Function	Failure Mode	Mechanisms and Cause of Failure	Effect of Failure	Current Controls	*Risk Ranking				Recommended Corrective Action	Actions Taken	Risk Ranking				Responsible Department
						P	S	D	R			P	S	D	R	
Mechanical Widget Housing Structure	Provides cable/indicator actuation	Indicator will not move freely along track	Incorrect tolerancing between indicator and base	Misregistration of gear select or inoperable	Installation procedure redesigned to install indicator flag opening using a flexible arm	6	10	1	60	Review DFMA study and process capability data regarding concerns relative to problem	Resolution due 7/26/94					J. Doe and A. Clark—Engineering and Manufacturing
		Cable not in correct path	Indicator has insufficient travel	Misregistration or inoperable	Control cable retention hook maximum dimension from cable guide	6	10	2	120	Increase cable retention hook maximum dimension to X mm	Cable retention increased to X dimension and retested per test plan—7/14/94	2	10	2	40	J. Doe—Engineering
		Cracked/broken mounting tabs	Insufficient wall thickness	Loss of function	Increase mounting tabs from X+/- Y mm thickness	4	10	1	40	None						

* Key

P = Probability (Chance) of Occurrence

S = Seriousness of Failure to System

D = Likelihood that Defect will Reach the Customer

R = Risk Priority Measure (P X S X D)

Figure 5.4a   Design Failure Mode and Effects Analysis (DFMEA)—Example

Subsystem Name: Widget
Application Year: 19XX
Supplier Responsibility: RST Engineering

Part Name	Function	Failure Mode	Mechanisms and Cause of Failure	Effect of Failure	Current Controls	*Risk Ranking				Recommended Corrective Action	Actions Taken	Risk Ranking				Responsible Department
						P	S	D	R			P	S	D	R	
Widget Circuit Board Lighting Asm.	Wave solder PCB lighting asm.	Insufficient or excessive solder	Incorrect machine settings	Loss of electrical continuity—no backlight to base structure	Maintain solder temperature (X +/- Y Deg. C)	6	7	3	126	Conduct DOE experiment #3 for determining optimal machine settings—3/29/XX	Solder temperature adjusted to Z +/- Y Deg. C—4/14/XX	2	7	3	42	L. Powers—Manufacturing and G. Smith—Quality Control
					Control conveyor speed (A +/- B mm/ min)	5	7	2	70	Adjust conveyor speed setting per DOE experiment #3	Conveyor speed decreased to C +/- B mm/min—4/14/XX	1	7	2	14	L. Powers—Manufacturing and G. Smith—Quality Control
					Maintain solder flux density (E +/- F kg/L)	3	7	1	21	Evaluate trend data with manufacturing and supplier—4/2/XX	None					D. Carey—Tooling L. Powers—Manufacturing
					Control over preheat temperature (R +/- T Deg. C)	3	7	3	63	Adjust pre-heat temp. according to DOE experiment #3. Closely monitor pre-heat temp. variation.	Pre-heat temp. maintained at same setting. Variable control chart ($\bar{X}$ and R) monitoring used.	2	7	3	42	L. Powers—Manufacturing and G. Smith—Quality Control

Potential Failure Mode	Potential Cause	Current Process Controls	P	S	D	R	Recommended Action	Action Taken and Results	P	S	D	R	Responsibility
Cold solder joints	Solder or solder flux contamination	Solder wave height (C +/- D mm)	5	7	3	105	Adjust solder wave height according to DOE experiment #3. Review equipment maintenance records.	Solder wave height adjuster repaired and tooled for automated control—4/18/XX	1	7	4	14	Same as above
	Loss of electrical continuity—no backlight	Visual aid—solder appearance	3	7	2	84	Consider automated visual inspection system. Evaluate cost-benefit.	Study to be completed on 4/24/XX					D. Carey—Tooling L. Powers—Manufacturing
		Cleaning procedure PS-XYZ	2	7	4	28	None						
		Visual aid—solder appearance	3	7	4	84	Consider automated visual inspection system. Evaluate cost-benefit.	Study to be completed on 4/24/XX					D. Carey—Tooling L. Powers—Manufacturing
		Pull test audit (min. Z Newtons)	2	7	3	62	Examine process control charts and determine appropriate sampling frequency-4/20/XX	Sampling frequency increased to 5 parts per 4 hr. time interval. Control variation.	2	7	2	28	G. Smith—Quality Control

* Key

P = Probability (Chance) of Occurrence

S = Seriousness of Failure to System

D = Likelihood that Defect will Reach the Customer

R = Risk Priority Measure (P X S X D)

Figure 5.4b Process Failure Mode and Effects Analysis (PFMEA)—Example

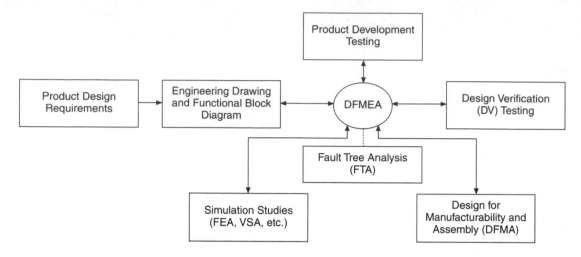

Figure 5.5   DFMEA and Its Relationship to Other Design Development Activities

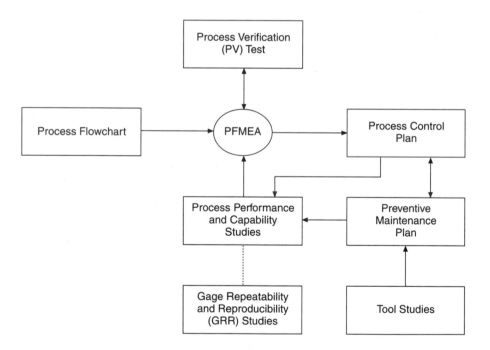

Figure 5.6   PFMEA and Its Relationship to Other Process Development Activities

B. Risk Analysis

2. Fault Tree Analysis (FTA)

What and Why the Tool Is Applied

The objective of a Fault Tree Analysis (FTA) is to determine the magnitude of risk and the critical path of those components or subsystems which effect a significant "top-event" failure condition. This fault tree combines both the qualitative and quantitative aspects of a component and its failure mode in order to present a pictorial representation of the fault path leading to the top-event failure. Fault tree analysis is a "top-down" as opposed to a "bottom-up" approach and it helps to see the interactive effects of components and processes that may impact the top-event. The goal in working with a fault tree analysis is to minimize the magnitude of risk, which may be contributed by the possible fault paths leading to the top-event failure condition. Fault tree analyses involve development of the possible fault paths, calculation of the risks associated with each possible path, and a prioritization of the most critical paths contributing to the top-event condition.

When the Tool Is Applied

Fault tree analyses are performed periodically throughout the design/process development phase in tandem with the FMEA. Usually the fault tree is applied when the development team decides that significant top-event subsystem or system condition exists and is affected by fault paths which could possibly contribute to the loss of a significant product performance function or cause a safety concern to the user of the product. The analysis may be repeated several times as components in the system are improved in order to reduce the potential of risk in the top-event condition.

Where the Tool Is Applied

Fault trees are applied to evaluate hardware designs and processes as well as software programs. Relative to hardware design, fault trees can be developed for evaluating: (1) how subcomponents influence top-event conditions of components, (2) how components influence top-event conditions for subsystems, and (3) how specific subsystems influence critical system operations. Relative to hardware manufacturing processes, fault trees can be used to analyze how particular process operations impact top-event conditions in the assembly of various product designs. On the other hand, fault tree analyses can be applied to software programs in order to help understand how particular subroutines control the potential for an unacceptable output response.

Who Is Responsible for Applying the Method

As was for the FMEA, it is the product design and process engineer who are the primary individuals responsible for helping to develop, update, and implement the results to improving their product or process. The product assurance engineer is a key team member of the team in facilitating the development and performing the analysis of the fault tree.

How the Tool Is Implemented

The fault tree for a hardware and software design or process essentially involves logic tree development, incorporation of probabilistic data to each event in the fault tree, analysis of the fault tree, and ranking of the most significant fault paths leading to the top-event condition.[2] An example of a completed fault tree and the ranking of those events that significantly influence the top-event is illustrated in Figures 5.7a and 5.7b.

The following highlights the procedure for conducting a fault tree analysis:

1. Develop the logic tree diagram of component, subsystem, or process failure modes, which interact with one another and lead to the occurrence of the identified top-event system condition. Use the functional block diagram (Figure 5.1) to help formulate the relationship of event failure modes to one another (see Figures 4.6 and 4.7).
2. Identify the probabilistic information for each event, which would include the failure mode apportionment and the unreliability of the component. The unreliability determined for each component in a design or process potential in a manufacturing operation should include a time element parameter to help correlate product performance to some product useful life period.

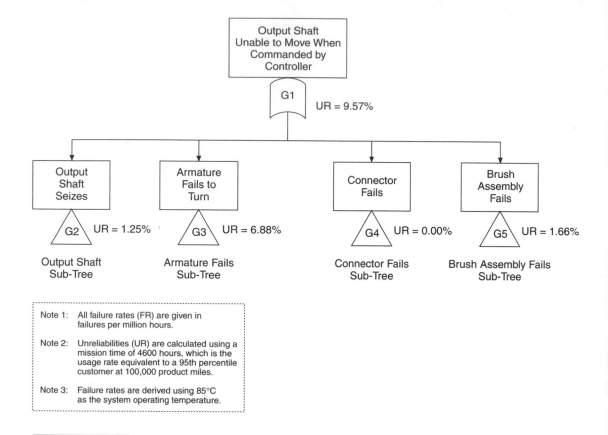

Source: John Bieda and Drew Hoelscher, "Comprehensive Design Reliability for the Automotive Component Industry via the Integration of Standard Reliability Methods," SAE Technical Paper Series, #910357, Feb. 1991, p. 4.

Figure 5.7a   Fault Tree Analysis—A DC Motor Assembly (Top-of-Tree Diagram) Example

3. Perform the analysis for each possible fault path and record the unreliability and importance parameter. The Fussel-Vesely importance measure is the fractional contribution of an event to the risk.[3] It is the ratio of the difference in the present risk and the decreased risk level with the feature optimized to the present risk. A high importance measure signifies a potential for the fault path of events to increase the risk in the top-event condition.
4. Rank order either the individual fault path probabilities and identify the highest fault path unreliability or importance (see Figure 5.8). The highest fault path unreliability or importance will identify the critical path leading to the top-event condition.
5. Determine the corrective action needed to decrease the importance of the events leading to the fault path, which increases the risk of the top-event.

Benefits of the Tool

1. Fault trees help to emphasize critical "top-event" failure. Customer dissatisfactions issues are explored. The technique forms the basis for risk reduction/cost benefit analysis.

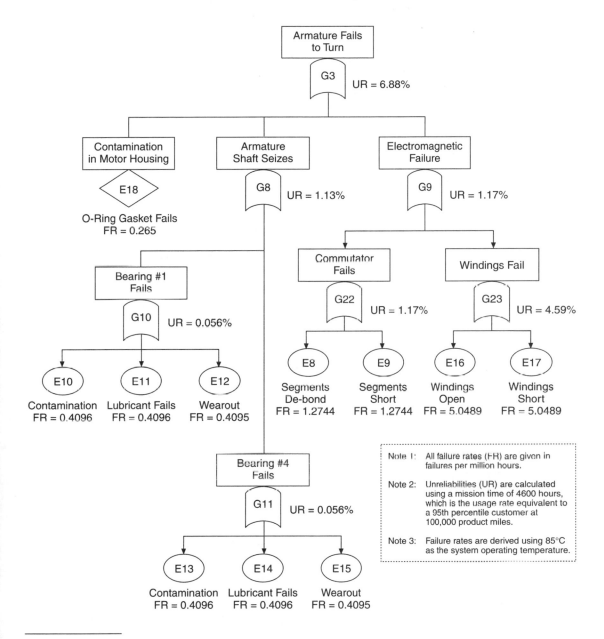

Source: John Bieda and Drew Hoelscher, "Comprehensive Design Reliability for the Automotive Component Industry via the Integration of Standard Reliability Methods," SAE Technical Series, #910357, Feb. 1991, p. 3. Reprinted with permission.

Figure 5.7b Fault Tree Analysis—A DC Motor Assembly (Armature Fails to Turn Sub-Tree) Example

2. Fault trees, when combined with the FMEA, help to produce a more complete product analysis method. The fault tree analysis contributes to sensitivity studies, severity issues, and risk evaluation.
3. Fault tree analyses help to combine the functional relationships of components and subsystems to yield system fault trees. This procedure helps illustrate the interactive effects of a system.
4. Fault tree analyses help provide valuable risk analysis information for a design review.

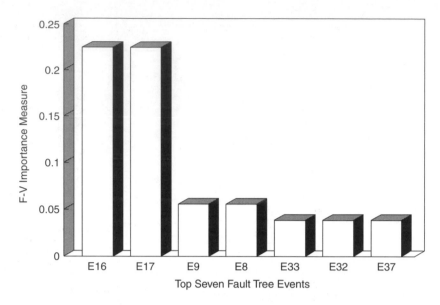

Figure 5.8 Fault Tree Analysis Fussel-Vesely Importance Measures—A DC Motor Assembly Example

C. Design for Manufacturability and Assembly (DFMA)

What and Why the Tool Is Applied

Design for Manufacturability and Assembly (DFMA) is a technique that quantifies the difficulties of assembly. The analysis spotlights not only parts that require excessive time for assembly, but also those that can be eliminated or combined with other parts. The term *Design for Manufacturability* includes a set of tools, each aimed at a particular aspect of the product design problem. The *Design for Assembly* procedure is just one of the tools in that kit, and is viewed as a vital starting point for the product development activity.

DFMA is used as a tool by the engineer to help consider an optimal design that focuses on the ease of assembly. It essentially involves a risk assessment of various assembly alternatives for various design considerations. An optimal design is chosen once the risk is sufficiently minimized as compared to other alternatives. Utilization of this tool helps to avoid costly mistakes in the manufacturability of a particular design during production. In addition, the DFMA tool stimulates creative thinking, builds teamwork, and provides benchmark measures for future evaluation.

When the Tool Is Applied

DFMA is applied early in the design and process development phases of the product development process. This technique is applied successively until the risk in the assembly of the product design has been minimized through the evaluation of several assembly alternatives.

Where the Tool Is Applied

DFMA is applied to any design level (component, subsystem, or system level) and part status (new, modified, or carryover).

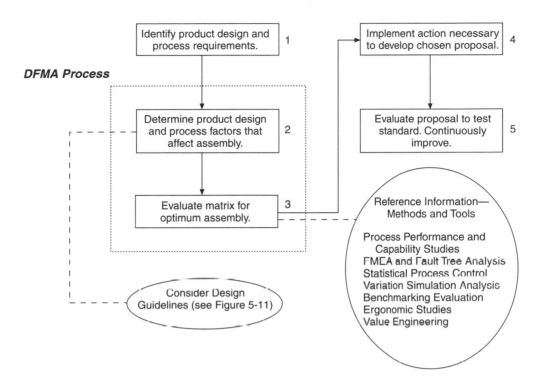

Figure 5.9   Descriptive Model for DFMA Evaluation Process

Who Is Responsible for the Method

Representatives for a DFMA would involve members from engineering, manufacturing, reliability and quality, product planning, and program management. All of these members would have input into reviewing the ease of assembly from many different viewpoints such as design, quality, reliability, cost, serviceability, maintainability, etc.

How the Tool Is Implemented

The DFMA process involves following a series of steps. These steps can be generally classified into the following five categories based on the traditional problem-solving model (see Figure 5.9):

1. Identify the product design and assembly system and all associated design and process requirements. Focus on the specific assembly issue of the design.
2. Determine all factors which may influence the decision of the product design's ability to be properly assembled. These factors should address characteristics of easy assembly, quality and reliability, serviceability, cost, maintainability, etc. There should be an agreement on a rating system, which would help to rank each product design proposal based on the ability to satisfy optimum levels of each of the identified factors. These ratings should be organized and ranked for each design proposal in a matrix format using a spreadsheet.

   Some general questions regarding the product design for assembly may involve the following:

   a. Do parts move relative to all other parts present?
   b. Must the parts be made of different materials?

Minimize the number of parts. Integrate parts.

The assembly process should proceed in a layered fashion (top to bottom)

Optimize part handling

Mating parts should be easy to align and insert

Choose one axis insertion or simple movements

Symmetrical parts simplify orientation for manual assembly or automatic feeding

Eliminate redundancies through value engineering

Exaggerate asymmetrical features of parts which must be nonsymmetrical

Reduce the fastener count and commonize screws. Avoid expensive fastening operations: i.e., threaded fasteners

Consider the production process capability

Avoid adjustments

Provide guide surfaces and datums for setup

Avoid tolerance demands and build variations

Evaluate assembly process for competitiveness

Avoid parts which tangle: i.e., coil springs

Source: Boothroyd, G. and P. Dewhurst, *Product Design for Assembly,* 1987.

Figure 5.10   Design for Manufacturing and Assembly—Guidelines to Good Design

    c. Would a combination of these parts prevent assembly or disassembly of other parts?
    d. Would servicing be significantly affected?

      Some specific guidelines to a good design relative to assembly considerations involve minimization of the number of parts, ease of part handling, symmetrical parts for simple orientation, mating parts which are easy to align and insert, nonadjustment capability, nonthreaded fasteners, etc.[4] See Figure 5.10 for DFMA guidelines as defined by Boothroyd-Dewhurst.

3. Evaluate the matrix and determine which product design proposal produces the most favorable score for the combined assessment of DFMA factors (see example in Figure 5.11). The evaluation should include supporting data from other analyses such as FMEA, fault trees, manufacturing performance and capability studies, benchmarking evaluations, variation simulation, design of experiment results, assembly ergonomic studies, etc.

4. Implement the action necessary to develop the chosen design proposal for easy assembly. Coordinate the team, assign tasks, and establish timing deadlines in order to carry out the decision.

5. Evaluate the proposed product design to a standard test, which helps validate the manufacturing processes ability to construct a robust product design.

Benefits of the Tool

1. DFMA helps stimulate creative thinking through the teamwork mechanism. It develops an awareness of and appreciation for a broader spectrum of concerns among the various disciplines involved.

2. DFMA provides benchmark measures of a design relative to an "ideal design."

3. DFMA encourages designers and engineers to see their product as envisioned by others, but to think about it differently.

4. DFMA provides a structured approach to design analysis by involving opportunities, evaluation, planning, and implementation.

Situation:

An investigation was necessary to review the opportunity of replacing manual fasteners with automated spot welding for a folding seat pivot assembly of a transportation device. A DFMA review was conducted during the product design/process development phase.

Results:

The following table was generated to review design/process alternatives per specific design/process considerations. The Boothroyd-Dewhurst method was used to derive design efficiency*, time, and cost:

DFMA Consideration	Design Alternative			
	A	B	C	D
Design Efficiency (%)	6.6	17.9	29.3	22.5
Assembly Time (sec.)	272.9	117.1	51.3	13.4
Assembly Cost ($)	1.091	0.468	0.205	0.053

Key:

Design A = Current design (benchmark)
Design B = Modified design including installation
Design C = Modified design
Design D = Installation using spot weld

* Design efficiency = (3 × total number of essential parts) / (total handbook assembly time)

Conclusion:

The results help to indicate that design alternatives C and D would be optimal choices based on the higher efficiency, short assembly time, and low assembly cost. It is up to the engineering team to further determine which of the two alternatives would be best. The inclusion of other design/process considerations would help the decision (process performance and capability results, ergonomics, etc.).

Figure 5.11   Design for Manufacturability and Assembly—Example

5. DFMA promotes product process design at the earliest possible stage of the product development process. It is at this very early stage in product development where implementation of guidelines to good design can be most beneficial from a cost-of-quality standpoint.
6. DFMA allows for concensus decision making.
7. DFMA helps in the evaluation and development of a good manual design. A good manual design must precede automated design in order to avoid complex, costly, and less productive assembly systems.
8. DFMA promotes the quest of minimizing the number of parts in order to simplify the assembly process as well as the design. This reduction in parts helps increase the quality and reliability of the product.

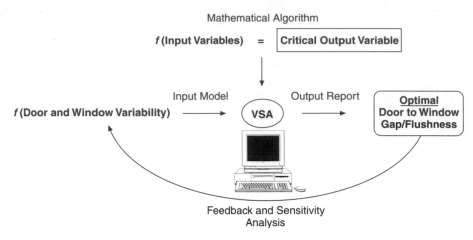

Figure 5.12 Variation Simulation Analysis Philosophy—An Example

D. Variation Simulation Analysis (VSA)

What and Why the Tool Is Applied

The purpose of variation simulation analysis (VSA) is to help predict what variation is present in a design given changes in particular design parameters. More specifically, VSA simulates a definable process by passing a set of imperfect components through an imperfect process to determine the variability of the resultant (see Figure 5.12). In this analysis process the engineering team develops a mathematical model to represent a particular engineering parameter and then simulates various inputs to the system. This process helps them understand the variation to the output parameter based on the interaction of inputs on the system. The execution of this analysis tool facilitates early design and process development of a product prior to hard tooling. In addition, this tool helps to evaluate questions posed during the design for manufacturability and assembly as well as address risks for an FMEA.

When the Tool Is Applied

Variation simulation analysis is most beneficial when applied during the latter stages of concept development and the early stages of product design and process development. The application of VSA can then be repeated as many times as it is necessary to minimize the variation in the desired output.

Where the Tool Is Applied

VSA is applied to any design or process complexity level (component, subsystem, or system level) and part status (new, modified, or carryover). This tool can be implemented on designs as well as processes. It can involve math models for mechanical, electrical, chemical, etc., engineering systems.

Who Is Responsible for the Method

VSA would be utilized mostly by the product design, process, or product releasing engineer during product design and process development. Detailed knowledge of the product design and process would also come from specialists of the engineering community such as the materials engineer, supplier, etc. The product assurance or quality/reliability engineer may assist the engineering team in the execution and interpretation of the results.

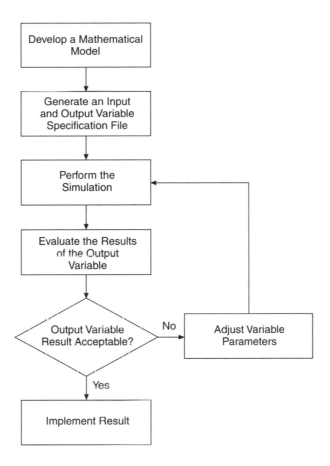

Figure 5.13 VSA Evaluation Process Procedure

How the Tool Is Implemented

This tool basically involves development of a mathematical model representing the output variable to be optimized, an input variable data file, simulation of the model, and an analysis of the variation as contributed by the input variables to the output variable (see Figure 5.13 and the example as illustrated in Figure 5.14). The use of a computer program such as Variation Simulation Modeling for the PC (VSM/PC) by Electronic Data Systems is helpful in conducting a variation simulation analysis. A summary of the steps leading to execution of this tool are as follows:

1. Develop a mathematical model representing the output variable as a function of input variables. Utilize engineering resources to facilitate and validate the model prior to simulation. Interface VSA with 3-D CAD system, structural analysis, mechanism analysis, manufacturing systems, and cost optimization technology.
2. Generate a data file of input parameter information, which would include nominal values and tolerances and a description of the input variable distribution. Identify the output variable specification limits for variable data analysis.
3. Execute the simulation of the program and model for a designated number of samples.
4. Produce an output report displaying descriptive statistics and a capability analysis relative to the sample distribution of the output variable. Evaluate the capability of the output parameter and the magnitude of input variable contributions to the output variable. At this point a decision is made to either accept the results and implement the change or adjust input variables and repeat simulation.

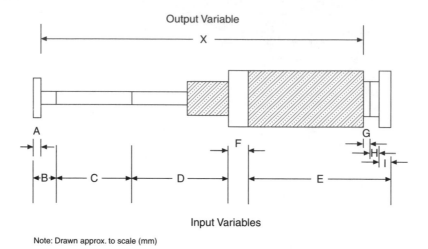

Figure 5.14a   Variation Simulation Analysis Example—Widget Assembly Drawing

```
C
      FUNCTION CALC ()
C     ROTOR SHAFT DESIGN SIMULATION
C     ROTOR SHAFT LENGTH STUDY
C
      IMPLICIT DOUBLE PRECISION (A-Z)
      COMMON/INPUT/DIMA,DIMB,DIMC,DIMD,DIME,DIMF,DIMG,DIMH,DIMI
      COMMON/OUTPUT/DIMX
      CALC=0
C
      DIMX=-DIMA+DIMB+DIMC+DIMD+DIME+DIMF-(DIMG+DIMH+DIMI)    ◄———  Math
      RETURN                                                        Model
      END
```

```
 File    Edit    Search    Options                                      Help
                               ─SHAFT.DAT─
C
      INPUT
      1 N 0.635 0.04 /*DIMA
      2 N 5.85 0.25 /*DIMB
      3 N 19.60 0.175 /*DIMC
      4 N 24.45 0.13 /*DIMD              ◄———  Input/Output
      5 N 37.71 0.13 /*DIME                     Data File
      6 N 5.00 0.13 /*DIMF
      7 N 1.50 0.05 /*DIMG
      8 N 3.11 0.10 /*DIMH
      9 N 3.555 0.13 /*DIMI
      OUTPUT
      1 DIMX,83.56,84.06,B,ROTOR SHAFT LENGTH




 MS-DOS Editor   (F1=Help) Press ALT to activate menus
```

Figure 5.14b   Variation Simulation Analysis Example—Mathematical Model and Input/Output Variable Specification File

```
PRESS <Enter> TO CONTINUE.....

    PANEL 3.1.3.1          VARIATION SIMULATION MODELING        VERSION 3.0
                              STATISTICS REPORT Panel



VARIABLE #  1 DIMX      B,ROTOR SHAFT LENGTH
SAMPLE SIZE     5000                        SIMULATED   STATISTICAL     LIMITS

NOMINAL         83.8100 LOW                 83.3149      83.3926      83.5600
MEAN            83.8102 HIGH                84.3467      84.2277      84.0600
STD DEV          0.1392 RANGE                1.0318       0.8351       0.5000
SKEWNESS         0.0157 CAPABILITY INDEX     0.4842       0.5983
KURTOSIS        -0.0382 SHIFT FROM NOM       0.0002

                        -------------------- % ANALYSIS --------------------

    DISTRIBUTION TYPE:  % BELOW LIMIT          3.7          3.6
       NORMAL (ASSUMED) % ABOVE LIMIT          3.8          3.6
                        % OUTSIDE LIMITS       7.5          7.2
                        % SPEC USED          206.4        167.0
                        % SHIFT FROM NOM       0.0

PRESS <Enter> TO CONTINUE.....


VARIABLE #    1 DIMX      B,ROTOR SHAFT LENGTH

                        HISTOGRAM    1'*' = 30 SAMPLES (SIMULATED DATA)
                                      '.' =  BEST FIT DISTRIBUTION
CUM
PROB   MIDPOINT    FREQUENCY

0.000     83.2000     0 +
0.000     83.3000     2 +
0.002     83.4000    19 +.
0.013     83.5000   136 +***.*
0.065     83.6000   457 +**************.-------------------------------------
0.214     83.7000  1072 +**********************************.*
0.471     83.8000  1378 +************^^^^*****************************.
0.741     83.9000  1140 +*************************************.
0.914     84.0000   575 +******************.
0.981     84.1000   188 +*****.------------------------------------------
0.997     84.2000    30 +.
1.000     84.3000     3 +
1.000     84.4000     0 +
1.000     84.5000     0 +
 PRESS <Enter> TO CONTINUE.....
```

Figure 5.14c Variation Simulation Analysis Example—Simulation Results for Original Condition

5. Adjust the nominal magnitude or tolerance limits of the significant input variable and repeat the simulation of the model if results are not acceptable, otherwise implement results. Strive for no out-of-tolerance conditions, parametric capability, and insignificant impact of input to the output variable variation.

Benefits of The Tool

1. VSA helps Engineering identify potential product and assembly problems early at the design development phase.
2. VSA facilitates the decision process of comparing alternate product and process designs.
3. VSA benefits the engineering team in helping to identify dimensional errors in part specifications.

```
    PANEL 3.1.4.5          VARIATION SIMULATION MODELING          VERSION 3.0
                                HLM REPORT Panel


    VARIABLE #    1  DIMX     B,ROTOR SHAFT LENGTH
                         NOMINAL      VARIANCE
          MAIN EFFECT    83.810     0.19429E-01

                                                              % OF OUTPUT RANGE
          INPUT                              % OF OUTPUT MAIN   BELOW     ABOVE
          VARIABLE        DESCRIPTION        EFFECT VARIATION  NOMINAL   NOMINAL

             2     DIMB                            35.8          50.0      50.0
             3     DIMC                            17.5          50.0      50.0
             4     DIMD                             9.7          50.0      50.0
             5     DIME                             9.7          50.0      50.0
             6     DIMF                             9.7          50.0      50.0
             9     DIMI                             9.7          50.0      50.0




    PRESS <Enter> TO CONTINUE.....


    PANEL 3.1.4.5          VARIATION SIMULATION MODELING          VERSION 3.0
                                HLM REPORT Panel

                                                              % OF OUTPUT RANGE
          INPUT                              % OF OUTPUT MAIN   BELOW     ABOVE
          VARIABLE        DESCRIPTION        EFFECT VARIATION  NOMINAL   NOMINAL

             8     DIMH                             5.7          50.0      50.0
             7     DIMG                             1.4          50.0      50.0

          1 VARIABLES CONTRIBUTE BELOW 1.0%      0.9
                                               ---------
                                                100.0




    PRESS <Enter> TO CONTINUE.....
```

Figure 5.14d Variation Simulation Analysis Example—Simulation Results for Original Condition

4. VSA provides the ability for Product Design and Process Engineering to evaluate proposed design or tooling changes.
5. VSA promotes improved design productivity and innovativeness.
6. VSA helps to reduce the number of problems and changes required after start of production.
7. VSA facilitates the reduction of internal cost of quality—rework and scrap.
8. VSA fits well with the ability to reduce lead times of program development programs.
9. VSA provides means to carryover knowledge.
10. VSA satisfies the goal of improved product quality, increased customer satisfaction, and reduced external costs of quality—warranty.

```
PANEL 3.1.3.1          VARIATION SIMULATION MODELING        VERSION 3.0
                         STATISTICS REPORT Panel

                                                            DIM B, F and I
                                                    ⟶      tolerances changed
                                                            to 0.05 mm

VARIABLE #  1 DIMX     B,ROTOR SHAFT LENGTH
SAMPLE SIZE    5000                      SIMULATED  STATISTICAL    LIMITS

NOMINAL      83.8100 LOW                    83.5799    83.6192   83.5600
MEAN         83.8103 HIGH                   84.0579    84.0013   84.0600
STD DEV       0.0637 RANGE                   0.4779     0.3821    0.5000
SKEWNESS      0.0225 CAPABILITY INDEX        1.0450     1.3070
KURTOSIS     -0.0349 SHIFT FROM NOM          0.0003

                    -------------------- % ANALYSIS -------------------

    DISTRIBUTION TYPE:   % BELOW LIMIT          0.0        0.0
      NORMAL (ASSUMED)   % ABOVE LIMIT          0.0        0.0
                         % OUTSIDE LIMITS       0.0        0.0
                         % SPEC USED           95.6       76.4
                         % SHIFT FROM NOM       0.1

PRESS <Enter> TO CONTINUE.....



VARIABLE #   1 DIMX      B,ROTOR SHAFT LENGTH

                    HISTOGRAM   1'*' = 26 SAMPLES (SIMULATED DATA)
                                '.' =  BEST FIT DISTRIBUTION
CUM
PROB   MIDPOINT    FREQUENCY

0.000      83.5200    0 +
0.000      83.5600    1 +-------------------------------------------
0.000      83.6000    3 +
0.004      83.6400   15 +*.
0.020      83.6800  170 +*****.*
0.078      83.7200  487 +*****************.*
0.215      83.7600  901 +*********************************.
0.436      83.8000 1199 +***********************************************.
0.680      83.8400 1117 +*********************************************.
0.863      88.8800  694 +**************************.
0.958      83.9200  303 +***********.*
0.991      83.9600   65 +**.
0.999      84.0000   22 +.
1.000      84.0400    3 +
1.000      84.0800    0 +-------------------------------------------------
1.000      84.1200    0 +
 PRESS <Enter> TO CONTINUE.....
```

Figure 5.14e Variation Simulation Analysis Example—Simulation Results for Revised Condition

E. Process Control Plan Development

What and Why the Tool Is Applied

The objective of process control plan development is to translate current controls as initially identified in the PFMEA to a scheduled control document, which would describe in detail how the product process control parameter is to be evaluated during the manufacturing process. This document is written to help address how key design/process characteristics are to be controlled during production. Each process operation from a manufacturing process, as illustrated in a process flowchart, is described in the process control plan. This document becomes input for the development of process instruction sheets which are used to help train operators in the execution and management of various pieces of equipment.

```
   PANEL 3.1.4.5           VARIATION SIMULATION MODELING         VERSION 3.0
                                HLM REPORT Panel



   VARIABLE #    1  DIMX    B,ROTOR SHAFT LENGTH
                       NOMINAL        VARIANCE
         MAIN EFFECT   83.810      0.40676E-02
                                                              % OF OUTPUT RANGE
         INPUT                           % OF OUTPUT MAIN     BELOW      ABOVE
         VARIABLE      DESCRIPTION       EFFECT VARIATION    NOMINAL    NOMINAL

            8    DIMH                          27.3           50.0       50.0
            7    DIMG                          27.3           50.0       50.0
            3    DIMC                           6.8           50.0       50.0
            4    DIMD                           6.8           50.0       50.0
            5    DIME                           6.8           50.0       50.0
            6    DIMF                           6.8           50.0       50.0




   PRESS <Enter> TO CONTINUE.....



   PANEL 3.1.4.5           VARIATION SIMULATION MODELING         VERSION 3.0
                                HLM REPORT Panel

                                                              % OF OUTPUT RANGE
         INPUT                           % OF OUTPUT MAIN     BELOW      ABOVE
         VARIABLE      DESCRIPTION       EFFECT VARIATION    NOMINAL    NOMINAL

            2    DIMB                           6.8           50.0       50.0
            9    DIMI                           6.8           50.0       50.0
            1    DIMA                           4.4           50.0       50.0

            0 VARIABLES CONTRIBUTE BELOW 1.0%    0.0
                                              ---------
                                                100.0
```

Figure 5.14f   Variation Simulation Analysis Example—Simulation Results for Revised Condition

When the Tool Is Applied

Process control plans are developed during the early stages of the product design and process development phase. The plan becomes a living document in that it is reviewed and modified as changes are incurred to a process. Any modifications to a process during product design/process development, validation, or production are incorporated through the process flowchart, PFMEA, and finally into the process control plan document. This document is used along with the process flowchart and PFMEA to form the logical sequence of a process description before operator instruction sheets are developed (see Figure 5.15).

Where the Tool Is Applied

Process control plans are applied to any design or process complexity level (component, subsystem, or system level) and part status (new, modified, or carryover).

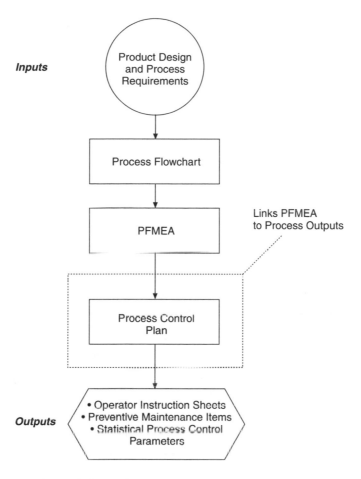

Figure 5.15 Process Control Plan Relationship to Other Process Development Activities

Who Is Responsible for the Method

The process engineer and tooling engineer are primary developers of the process control plans. The product releasing engineer, design engineer, and product assurance specialist are important members of the team as far as the review of the document. Product Assurance helps to assure that the document adequately addresses and develops principal controls of key product/process characteristics for specific manufacturing operations.

How the Tool Is Implemented

Process control plan development occurs once the process has been defined and a risk analysis performed. From the initial process description and risk analysis, the process control plan helps highlight specific description of process operation and equipment, key product design/process parameters, specification of the parameter, measurement evaluation technique, sampling plan, control method, and reaction plan. If the process operation is modified as a result of a requirement change, test results, etc., then the specific operation affected would be adjusted to reflect the revised condition. A brief step-by-step description of the construction of a process control plan is as follows (see Figure 5.16)[5]:

1. Label pertinent information about the product such as part number, manufacturing location, team members, responsible parties, date, etc.

Supplier   ATB Corporation
Control Plan #   123
Part Description   Widget Electronic Circuit Bd. Asm.
Part Number   CB12345

Part/Process Number	Process Name or Operation Description	Machine, Device, Jig Tools for Mfg.	No.	Characteristics		Special Char. Class
				Product	Process	
3–4	Manual Insertion	Automatic Parts Feeder (ABC Manufacturing)		Insertion		
				Value		
				Polarity		
					ESD Control	
					PCB Handling	
4–1	Spray Fluxor	Product Fluxor (XYZ Manufacturer)		Clean pads before soldering		
				Proper PCB		
					Flux Density	
					Conveyor Speed	
					Fluxor Filter Cleaning	
					Machine Air Pressure	
					Flux Motor Rotation	
					Conveyor Width	
4–2	Wave Solder	Soldering Machine –Jig #1 for soldering main PCB, Jig #2 for soldering daughter boards, Jig #3 for soldering front PCB			Preheat temperature	
					Solder Temp.	
					Conveyor Speed	
					Solder Quality	
					Angluar Position Wave Height 1 Wave Height 2	
					Solder Composition	
					Conformity	

Figure 5.16   Process Control Plan—Electronic Circuit Board Manufacturing Process Example

Product/Process Special or Tolerance	Evaluation Measurement Technique	Sample		Control Method	Reaction Plan
		Size	Frequency		
Install right part number per work station	Visual identification	100%	Continuous	In-circuit test P Chart	Apply Procedure #4
Same as above	Same as above	100%	Continuous	In circuit test P Chart	Stop process and inform technician Perform RCA #3
Per instruction sheet #4	Same as above	100%	Continuous	In circuit test P-Chart	
Apply instruction sht. #10	ESD Meter #X23	3	Every 4 hrs.	Same as above	
Use ESD racks and gloves				P-Chart	
Instruction Sheet #23					
Automatic Inspection Code					
850 +/– 10 mg per cubic cm	Densimeter #1	4	Every 8 hrs.	X-bar and R Chart	Stop process and inform technician Perform RCA #3
0.9 +/– 0.1 m per min.	Measurement Datapack Digitizer	1	Every 8 hrs.	P-Chart	
Instruction Sheet #3	Visual	1	Every 8 hrs.	Log	
1.5 KGF per square cm	Machine Manometer	2	Every 4 hrs.	X-bar and R Chart	
2.0 +/– 0.2 m per min.	Measurement Datapack Digitizer	2	Every 8 hrs.	X-bar and R Chart	
24.8 +/– 0.2°	Soldering Jig #A1	2	Every 8 hrs.	X-bar and R Chart	
340 +/– 10° C	Digitizer Machine #1	4	Every 4 hrs.	X-bar and R Chart	Stop process and inform technician Perform RCA #3
240 +/– 5° C	Temperature Meter	4	Every 4 hrs.	X-bar and R Chart	
1000 +/– 50 mm per min.	Machine Digitizer	4	Every 4 hrs.	X-bar and R Chart	
Instruction Sheet #45	Measurement Datapack Digitizer	3	Every 8 hrs.	Standard Dev. Chart	
6 +/– 1 angular°		4	Monthly	X-bar and R Chart	
15 +/– 2mm	Glass Method	3	Every 8 hrs.	X-bar and R Chart	
20 +/– 5 mm	Same as above	3	Every 8 hrs.	X-bar and R Chart	
min. ratio Cu 0.005%					
QP9000 Standard	Visual	100%	Continuous	Log	

2. Identify each process step number and provide a brief description of operation and equipment.
3. Determine the key product or process characteristic for that particular operation and identify its importance.
4. Identify the tolerance or specification limits for the particular product design and process characteristic.
5. Determine the appropriate measurement instrument or evaluation technique to be used to measure the particular characteristic. This instrument or technique will be subject to calibration, gauge repeatability, and reproducibility analysis.
6. Select the sample size and frequency to be taken in order to determine process control.
7. Identify the control method to be used to monitor and evaluate the process control relative to the key characteristic.
8. Describe the reaction plan to be implemented in the event of an out-of-control event for the key characteristic under observation.

Benefits of the Tool

1. Process control plans help to reduce waste and improve the quality of products during design, manufacturing, and assembly.
2. Process control plans provide a structured approach to facilitate the evaluation of a product and process.
3. Process control plans help to identify each process step's key characteristics and measurement methods necessary to evaluate the process capability.
4. Process control plans help to focus resources on processes related to characteristics that are important to the customer.
5. Process control plans function as the key control document, which takes input from the PFMEA and establishes the current controls for proper evaluation.
6. Process control plans help to communicate changes in the product/process to the engineering and manufacturing community.
7. Process control plans function as the central process description document that aids in the development of operator instruction sheets.
8. Process control plans help establish preventive maintenance plans by identifying important process control variables, which may need proper adjustment in order to maintain process control.

F. Measurement System Evaluation

What and Why the Tool Is Applied

The principal objective of measurement system evaluation is to determine the capability of a measurement tool, gauge, or operator system. This evaluation process involves assessment of a measurement system's calibration and gauge repeatability and reproducibility. Assessment of a measurement system is necessary in order to better understand the source of variation in a piece of equipment or inspection system and avoid misinterpretation of whether the measurement instrument or product design/process is responsible. Assuring that the measurement system is capable involves evaluating the calibration by comparing the instrument or process to an acceptable standard and then evaluating the gauge repeatability and reproducibility by determining the total error as a function of part variation and operator variation. Measurement system evaluation provides the basis for gauge stability over time, repair and recertification, control, benchmarking standards (master), etc.

When the Tool Is Applied

Measurement system evaluation is applied continuously throughout a product's development cycle, including production. The assessment of measurement capability should be performed during a prescribed time interval and especially when the product design or manufacturing process capability is in question. Calibration should always precede gauge repeatability and reproducibility in order to determine gross error of the instrument to the master standard. Gauge repeatability and reproducibility should then follow calibration in order to locate the source of variation and promote improvement of the measurement system before its use.

Where the Tool Is Applied

Measurement system evaluation is applied to any measurement tool, gauge, or inspection system (including visual control). The analysis is performed on any part condition (new, modified, or carryover) and is repeated at some designated frequency throughout the product development process, including production.

Who Is Responsible for the Method

Responsible personnel for implementation of measurement system evaluation primarily involve the process engineer and manufacturing engineer. Other team members who may contribute to gauge improvements or alternate measurement approaches include the product assurance specialist, product release engineer, tooling engineer, and supplier quality specialist.

How the Tool Is Implemented

Measurement system evaluation is accomplished by first determining the calibration status and then analyzing the gauge repeatability and reproducibility of the particular measurement system.

Calibration

Calibration of a tool, gauge, piece of equipment, or inspection system that includes variable data involves determining the accuracy as a function of the percent tolerance or process variation. The calibration of a measurement system that includes attribute data (that is, go/no-go, etc.) would involve the comparison of each measured or inspected part to some standard and a determination of acceptability. The specific steps necessary for determining the calibration status of a piece of equipment or inspection system is highlighted as follows:

1. Make measurements or visually evaluate an appropriate number of samples. If samples involve physical measurement, then record the variable data. If samples are visually inspected, then compare sample to a master standard for acceptability and record status. The master value can be determined by averaging several measurements with the most accurate measuring equipment available.
2. Determine the average for the samples measured.
3. Calculate the accuracy for the variable data measurement system by taking the difference between the observed average of measurements and the master value:

$$\text{Accuracy} = \text{Master Measurement} - \text{Observed Average} \qquad (13)$$

   The percent accuracy can be calculated by dividing the difference in the observed average value and the master value (accuracy) by the tolerance or process variation:

$$\text{Percent Accuracy} = (\text{Accuracy/Tolerance}) \times 100 \qquad (14)$$

See the example in Figure 5.17 for an application of the calibration accuracy analysis process.

The accuracy is determined by the difference between the master measurement and the observed average measurement. To accomplish this, a sample of one part is measured ten times by one operator. The values of the ten measurements are listed below. The master measurement determined by layout inspection equipment is 0.80 mm and the process variation for the part is 0.70 mm.

$$
\begin{array}{ll}
X_1 = 0.75 & X_6 = 0.80 \\
X_2 = 0.75 & X_7 = 0.75 \\
X_3 = 0.80 & X_8 = 0.75 \\
X_4 = 0.80 & X_9 = 0.75 \\
X_5 = 0.65 & X_{10} = 0.70
\end{array}
$$

The observed average is the sum of the measurements divided by 10.

$$
\overline{X} = \frac{\Sigma X}{10} = \frac{7.5}{10} = 0.75
$$

The accuracy is the difference between master measurement and observed average,

Accuracy = Master Measurement – Observed Average
Accuracy = 0.80 – 0.75 = 0.05

The percent of process variation for accuracy is calculated as follows:

% Accuracy = 100 [Accuracy/Process Variation]
% Accuracy = 100 [0.05/0.70] = 7.1%

The percent of tolerance for accuracy is calculated in the same way, where tolerance is substituted for process variation.

Therefore, the thickness gauge to be used in the gauge R&R study has an accuracy of 0.05 mm. This means that the observed measurements on the average will be 0.05 mm from the master measurement and this accuracy will consume 7.1% of the process variation.

Source: Chrysler, Ford, and General Motors Supplier Quality Requirments Task Force, *Measurement Systems Analysis Reference Manual.* 1990, Section 2, pp. 27–28. Reprinted with permission.

Figure 5.17   Measurement System Calibration Analysis—Measurement Accuracy Example

Gauge Repeatability and Reproducibility (GRR)–Variable Data Measurement Systems

Variable data measurement systems offer the capability of evaluating the magnitude of the total measurement error. Variable measurement system analysis is very informative to the engineer, because of the ability to measure the degree of variation between the measured part and the specification limit. Gauge repeatability and reproducibility analysis attempts to further evaluate the variation in the measurement system based on the variation between parts and operators. Specifically, gauge repeatability refers to the variation in measurements obtained with one variable type data gauge when used several times by one operator while measuring the identical characteristics on the same parts. Gauge reproducibility refers to the variation in the average of the measurements made by different operators using the same gauge when measuring characteristics on the same parts. The GRR study is performed using either variable or attribute type data.

A GRR study for variable type data can be performed using the *range method* (short-form approach) or *average and range method* (long-form approach). The range method essentially consists of determining a total GRR as either a function of part tolerance or process variation.[6] It does not further evaluate the contribution to the total variation. This range method involves: (1) a minimum of five samples, (2) two operators, (3) one variable data type gauge, and (4) the gauge being in calibration prior to the study. The goal is to have a total GRR as small as possible. A sample of the method is found in Figure 5.18.

The average and range essentially consists of determining a total GRR as either a function of part tolerance or process variation and includes the contribution of part and operator variation to the total variation.[7] The conditions surrounding this approach involve: (1) a sample size of 10 parts, (2) three operators, (3) one variable data type gauge, (4) at least two trials per operator, and (5) the gauge being in calibration before the study. The goal is to have a total GRR as small as possible (for example, less than 10% is acceptable, 10%–30% is marginal, and greater than 30% is unacceptable). A basic procedure in conducting the study is as follows (see Figure 5.19):

1. Measure 10 parts at least two times for all three operators involved in the investigation. Ensure that the part selection process is randomized in order to prevent biasing the study.
2. Compute the average and range for each part and each operator.
3. Determine the average of the averages by part and operator $(\bar{\bar{X}})$. Determine the average of the ranges by part and operator $(\bar{\bar{R}})$.
4. Subtract the smallest part average from the largest part average and enter the result for R_p. Calculate the average of the $\bar{R}$'s for each operator.
5. Determine the equipment variation (*EV*) and the appraiser variation (*AV*). The constants K_1, K_2, and K_3 can be determined using Table 5a:

$$EV = \bar{R} \times K_1 \tag{15}$$

where: EV = equipment variation
   $\bar{R}$ = average range
   K_1 = a constant which depends on the number of trials used in study

$$AV = \sqrt{\left(\bar{X}_{Diff} \times K_2\right)^2 - \left(\frac{EV^2}{nr}\right)} \tag{16}$$

where: AV = appraiser variation
   $\bar{X}_{Diff}$ = maximum average operator difference
   K_2 = a constant which depends on the number of operators in the study
   EV = equipment variation
   n = number of parts
   r = number of trials

6. Compute the repeatability and reproducibility (*RR*) by taking the square root of the sum of the squares for *EV* and *AV*:

$$RR = \sqrt{\left(EV^2 + AV^2\right)} \tag{17}$$

7. Calculate the part variation (*PV*), which is based on R_p and a constant K_3.

$$PV = R_p \times K_3 \tag{18}$$

where: PV = part variation
   R_p = range of the part averages
   K_3 = a constant based on the number of parts in the study

Part No. and Name: Housing
Characteristics: Width
Specification: 10.0–10.6 mm

Gage Name: Micrometer
Gage No: G-34
Gage Type: 0.00–100.00 mm

Date: 1/30/97
Performed by: J. Doe

	Part 1	Part 2	Part 3	Part 4	Part 5
Operator A	10.300	10.100	10.220	10.500	10.400
Operator B	10.300	10.080	10.150	10.530	10.390

Enter Total Tolerance Range ⟶ 0.600
(. . . or enter Process Variation if desired)

	Range Part 1	Range Part 2	Range Part 3	Range Part 4	Range Part 5
Part Range	0.000	0.020	0.070	0.030	0.010

Average Range: 0.026

GR&R = 4.33 (Average Range)

GR&R: 0.113

% GR&R = 100[GR&R / Total Tolerance] or. . .
        = 100[GR&R / Process Variation]

% GR&R: 18.8%

Acceptable: under 10%   **Marginal: 10% to 30%**   **Unacceptable: over 30%**

Source: Chrysler Corporation, *Process Sign-off Companion Training Manual*, January 1996, p. 121. The methodology used in this worksheet format as developed by Chrysler was obtained from the Ford, General Motors, Chrysler *Measurement Systems Analysis Reference Manual* as distributed by AIAG, Section 4, p. 38.

Figure 5.18 Variable Measurement Systems Repeatability and Reproducibility Study—Range Method (Short Method) Example

Part No. and Name: Housing
Characteristics: Width
Specification: 10.0–10.6 mm

Gage Name: Micrometer
Gage No: G-34
Gage Type: 0.00–100.00 mm
Date: 1/30/97
Performed by: J. Doe

Operator A

	Part 1	Part 2	Part 3	Part 4	Part 5	Part 6	Part 7	Part 8	Part 9	Part 10	Average
1	10.650	10.200	10.550	10.350	10.100	10.000	10.450	10.450	10.100	10.250	10.310
2	10.600	10.200	10.450	10.300	10.000	10.000	10.400	10.500	10.100	10.150	10.270
3											
Avg	10.625	10.200	10.500	10.325	10.050	10.000	10.425	10.475	10.100	10.200	$\bar{X}_A$ = 10.290
Rng	0.050	0.000	0.100	0.050	0.100	0.000	0.050	0.050	0.000	0.100	$\bar{R}_A$ = 0.050

Operator B

	Part 1	Part 2	Part 3	Part 4	Part 5	Part 6	Part 7	Part 8	Part 9	Part 10	Average
1	10.550	10.150	10.400	10.300	10.050	9.950	10.450	10.500	10.100	10.250	10.270
2	10.600	10.100	10.500	10.300	10.050	10.000	10.450	10.500	10.000	10.150	10.265
3											
Avg	10.575	10.125	10.450	10.300	10.050	9.975	10.450	10.500	10.050	10.200	$\bar{X}_B$ = 10.268
Rng	0.050	0.050	0.100	0.000	0.000	0.050	0.000	0.000	0.100	0.100	$\bar{R}_B$ = 0.045

Operator C

	Part 1	Part 2	Part 3	Part 4	Part 5	Part 6	Part 7	Part 8	Part 9	Part 10	Average
1	10.600	10.250	10.450	10.400	10.100	10.050	10.400	10.600	10.200	10.150	10.320
2	10.600	10.200	10.500	10.300	10.050	10.050	10.350	10.500	10.150	10.150	10.285
3											
Avg	10.600	10.225	10.475	10.350	10.075	10.050	10.375	10.550	10.175	10.150	$\bar{X}_C$ = 10.303
Rng	0.000	0.050	0.050	0.100	0.050	0.000	0.050	0.100	0.050	0.000	$\bar{R}_C$ = 0.045

| Part Avg | 10.600 | 10.183 | 10.475 | 10.325 | 10.058 | 10.008 | 10.417 | 10.508 | 10.108 | 10.183 | Part R_p = 0.542 |

Part $\bar{R}_{dbar}$ = 0.047
$\bar{X}_{diff}$ = 0.035
UCL_r = 0.153
LCL_r = 0.000

(If R&R as % of Tolerance is Reqd.)
Enter the Total Tolerance Range Here → 0.600

Process Variation, TV (Determined from Sample Values) → 0.907

Gage R & R

		Percent of Tolerance	Percent of Process
Equipment Variation =	0.213	35.5%	23.5%
Appraiser Variation =	0.082	13.6%	9.0%
Part Variation =	0.878	146.3%	96.3%
Repeatability and Reproducibility =	0.228	38.0%	25.1%

Acceptable: under 10%
Marginal: 10% to 30%
Unacceptable: over 30%

Source: Chrysler Corporation, *Process Sign-off Companion Training Manual*, January 1996, p. 111. The methodology used in this worksheet as developed by Chrysler was obtained from the Ford, General Motors, Chrysler *Measurement Systems Analysis Reference Manual* as distributed by AIAG, Section 4, pp. 43–44. The constants used in calculating the R&R were obtained from "Quality Control and Industrial Statistics" by A. J. Duncan.[8]

Figure 5.19 Variable Measurement Systems Repeatability and Reproducibility Study—Average and Range Method (Long Method) Example

TABLE 5a

Average and Range Method Constants

Trials	K_1		Parts	K_3
2	4.56		2	3.65
3	3.05		3	2.70
			4	2.30
			5	2.08
			6	1.93
			7	1.82
			8	1.74
			9	1.67
			10	1.62

Operators	2	3
K_2	3.65	2.70

All calculations are based upon predicting 5.15 sigma (99.0% of the area under the normal distribution curve). K_1 is $5.15/d_2$, where d_2 is dependent on the number of trials m, and the number of parts times the number of operators g, which is assumed to be greater than 15. Regarding AV, if a negative value is calculated under the square root sign, the appraiser variation (AV) defaults to zero (0). K_2 is $5.15/d_2$, where d_2 is dependent on the number of operators m and g is 1, since there is only one range calculation. K_3 is $5.15/d_2$, where d_2 is dependent on the number of parts m and g is 1, since there is only one range calculation. d_2 is obtained from Table D_3, "Quality control and Industrial Statistics." A. J. Duncan.[8]

8. Determine the total variation (TV), which is a function of the computed RR and PV.

$$TV = \sqrt{\left(RR^2 + PV^2\right)} \tag{19}$$

9. Calculate the percent RR from the PV and TV.

$$\%RR = 100 \times \left(\frac{PV}{TV}\right) \tag{20}$$

When evaluating GRR as a percentage of the tolerance, the method of specifying limits will influence the outcome. The easiest case is two-sided tolerancing where upper and lower tolerance limits are specified. In this instance the Average and Range Method described earlier should be used. For a lower one-sided situation it is advisable to use percentage of process variation or the Average and Range—Control Chart Methodology (see *Measurement Systems Analysis* manual, section four). For an upper one-sided situation assume the lower limit is zero and use the Average and Range Method, or use the Average and Range—Control Chart Methodology.

Gauge Repeatability and Reproducibility (GRR)–Attribute Data Measurement Systems

An attribute data measurement system is evaluated differently from a variable data measurement system. The attribute gauge usually involves a comparison of a particular part to a very specific set of limits in order to either accept or reject the part. Attribute gauges are typically set up to check the part for conformance to the specification and cannot yield variable type data regarding the magnitude of conformance or nonconformance. Given the nature of the attribute gauge system a repeatability and reproducibility study will yield information of gauge measurement capability from a qualitative perspective.

The gauge repeatability and reproducibility study involving attribute data can be performed using a very simple technique. The conditions for conducting such an analysis involve: (1) selecting 20 parts such that at least two parts are above and two parts are below the specification limits, (2) choosing two operators, (3) using a pass/fail gauge or visual inspection system, and (4) ensur-

| Sample | Operator A | | Operator B | |
Number	1	2	1	2
1	Good	Good	Good	Good
2	No Good	No Good	No Good	No Good
3	Good	Good	Good	Good
4	Good	Good	Good	Good
5	Good	Good	No Good	No Good
6	No Good	Good	No Good	No Good
7	Good	Good	Good	Good
8	No Good	No Good	No Good	No Good
9	Good	Good	Good	Good
10	Good	Good	Good	Good
11	Good	Good	Good	Good
12	Good	Good	No Good	Good
13	No Good	No Good	No Good	No Good
14	No Good	No Good	No Good	No Good
15	Good	Good	Good	Good
16	No Good	No Good	No Good	No Good
17	Good	Good	Good	Good
18	No Good	No Good	No Good	No Good
19	Good	Good	Good	Good
20	Good	Good	Good	Good

Directions: Assure that at least two samples are chosen above and below the specification limit or do not meet the visual inspection criteria. An acceptable measurement system is obtained when all sample evaluations are consistent between operators and within trial inspections for an operator.

Analysis: Sample numbers 5, 6, and 12 did not produce consistent results between operators A and B, therefore, the measurement system is not acceptable.

Source: Chrysler, Ford, and General Motors Supplier Quality Requirements Task Force, *Measurement Systems Analysis Reference Manual,* 1990. Section 6, p. 79. Reprinted with permission.

Figure 5.20 Attribute Measurement Systems Repeatability and Reproducibility Study—An Example

ing the gauge is in calibration prior to the study. The steps for conducting such an analysis are as follows (see Figure 5.20).[9]

1. Measure and record all parts twice per operator.
2. Determine the pass or fail decision for each part.

Acceptance of the gauge or visual measurement system would occur if all measurement decisions agree (four per part). A nonacceptance would occur if all measurements did not agree. If an unacceptable study is determined, then improvement or a new measurement system must be pursued.

Benefits of the Tool

1. Measurement systems analysis provides the ability to assess the variability of a measurement system and allow opportunity to improve the system prior to its use in verifying product or process quality.
2. Measurement systems analysis helps in the understanding of the source and magnitude of total variation present in a given measurement system. It helps understand the contribution of equipment and appraiser variation to the total variation.

3. Measurement systems analysis provides methods for evaluating variable and attribute type data measurement systems.

4. Measurement systems analysis involves a statistical approach to the assessment of measurement systems in order to understand the risk level of misclassifying any product tested. Factors such as the statistical capability of the gauge to measure parts, cost, ease of use, etc., all become part of the decision-making process for the most appropriate gauge or measurement system.

5. Measurement systems analysis allows for improvement in the ability to discriminate between good and bad product and avoid the possibility of rejecting a bad part when it is good or accepting a good part when it is bad (Type I and Type II error).

G. Preventive Maintenance

What and Why the Tool Is Applied

Preventive maintenance involves a scheduled list of activities to ensure stability of a machine, test stands, and gauges which are fundamental to the repeatability of production build and measurement capability. Maintainability can be expressed as the probability that a machine can be retained in, or restored to, operational readiness within a specified interval of time when maintenance is performed according to specified instructions. It is this maintenance activity, which is necessary to assure stability of an operation, that is key to preventive maintenance. The implementation of preventive maintenance for the manufacture of a product is a key ingredient in maintaining production efficiency, ensuring process control, and striving for lower total life cycle or quality costs for an organization.

Major areas of maintenance include scheduled maintenance, and tool and equipment studies. Scheduled maintenance primarily involves tests and checkouts, scheduled servicing, periodic calibrations, planned overhauls, and replacement of worn parts. Tool and equipment studies essentially involve life expectancy and repair data analyses to determine the interval necessary for particular maintenance actions.

When the Tool Is Applied

Preventive maintenance development occurs throughout the product design and process development phases of the PDP. It begins with the evaluation of tools and equipment through life expectancy studies. It finishes with the development of specific preventive maintenance schedules as a result of the identification of key process control parameters for a specific machine or manufacturing operation through the process control plan. These key process controls become elements for the development of preventive maintenance instructions, which are carried out based on the preventive maintenance schedule as used during the production phase.

Where the Tool Is Applied

Preventive maintenance is applied on all product design levels (component, subsystem, or system level) and part status (new, modified, or carryover).

Who Is Responsible for the Method

The manufacturing, process, and tooling engineers are the primary people responsible for the development and implementation of preventive maintenance activities. The tooling engineer is responsible for the development and verification of equipment and tools for the process. He must design the tool, produce, and evaluate the life expectancy of the tool for the designated operation. This involves tool and equipment studies to measure the time between repair. The manufacturing and process engineers must develop and verify the complete operation of the manufacturing

Problem:

Determine the maintainability for a particular piece of equipment given a mean time to repair (MTTR) of 2 hours and a repair rate which is exponentially distributed. Characterize the equipment maintainability for several times to repair.

Solution:

1. Apply the exponential equation form for maintainability as a function of the time to restore:

$$M(t) = 1 - e^{-\frac{t}{MTTR}}$$

where: $M(t)$ = Maintainability
t = Time to restore (hours)
$MTTR$ = Mean Time to Repair (hours)

Express the maintainability equation relative to known information from problem:

$$M(t) = 1 - e^{-\frac{t}{2}}$$

2. Calculate the maintainability per various times to repair using the maintainability equation.

Time to Repair (hours)	Maintainability $M(t)$
0.5	0.22
2	0.63
4	0.87
6	0.95
8	0.98
10	0.99

Figure 5.21   Maintainability Study Example

sequence for the product to be produced. This may involve configuring the equipment in a assembly sequence, evaluating the characteristics of each operation in the process, and establishing the measurement tools and devices for verifying equipment capability. The product assurance specialist aids in the review and analysis of these tool and equipment studies and preventive maintenance plans. His or her role is to help assure that the tool and equipment studies are done properly and are accurately reflected in the scheduled preventive plan.

How the Tool Is Implemented

Preventive maintenance development and analysis is performed through tool and equipment studies and a scheduled preventive maintenance plan. Tool and equipment studies specifically involve evaluation of the life expectancy and repair data of the pieces of equipment used in the manufacturing process in order to define the appropriate time interval for maintenance action. One of the approaches and steps for conducting a study is briefly mentioned in the following list (see example in Figure 5.21).[10]

Tool and Equipment Life Studies

1. Determine the mean time to correctively repair, replace, or restore the equipment or tool to a satisfactory condition or operational readiness (MTTR).

2. Identify the mathematical distribution of the duration of repair, replacement, or restoration data. Express the time-to-restore distribution equation in terms of MTTR. Transpose the MTTR to a repair rate and rewrite the equation. This should simplify the calculation process to be described later.

    For example, using the probability density function for an exponential time-to-restore distribution equation in terms of the MTTR:

$$g(t) = \left(\frac{1}{MTTR}\right) e^{-\frac{t}{MTTR}} \tag{21}$$

where: $g(t)$ = probability density function
$MTTR$ = mean time to repair
t = time to restore

Utilizing the relationship of repair rate $u(t)$ to the MTTR:

$$u(t) = \frac{1}{MTTR} \tag{22}$$

Maintainability $M(t)$ can be determined once the mathematical integration is performed:

$$M(t) = 1 - e^{-\frac{t}{MTTR}} \tag{23}$$

3. Calculate maintainability or time to restore:

    Solve for the maintainability $M(t)$ of the equipment or tool for various times to restore. Examine the point of interest on the graph of maintainability $M(t)$ versus the time to restore (t)—see example in Figure 5.22.).

    Calculate the time to restore (t) based on different levels of maintainability [$M(t)$ expressed in percent].

    For example, in the exponential time-to-restore distribution case:

$$t = \frac{-ln(1 - M(t))}{u} \tag{24}$$

where: t = time to restore
$M(t)$ = maintainability (%)
u = repair rate (repairs/time unit)

4. Utilize the results of the time to restore the equipment for the scheduled maintenance plan.

Scheduled Preventive Maintenance Plan

The other consideration in preventive maintenance (PM) is development of the scheduled maintenance plan. This plan contains a schedule of the interval of time in which a specific maintenance action such as a repair, replacement, or restoration is performed on a piece of equipment for a particular process assembly operation. The principal objective is to define the maintenance action at the correct interval of time so that the tendency for the equipment to produce significant variation in output, or fail and produce significant downtime, is reduced

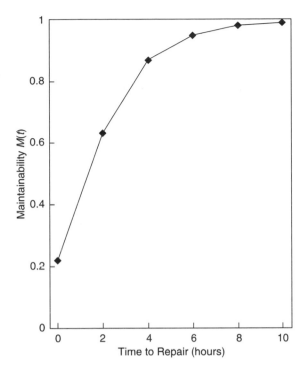

Source: Dr. Dimitri Kececioglu, PE, *Lecture Notes on AME 508-Advanced Reliability Engineering,* 1986, Chapter 6, p. 18. Adapted by permission.

Figure 5.22   Maintainability versus Time to Repair—An Example

and avoided. The following description helps to understand the construction of the PM plan (see example in Figure 5.23):

1. Identify the process number and piece of equipment involved in the particular manufacturing process. This identification scheme should correlate to the operations as illustrated in the process flow diagram.
2. Describe the key design or process parameter which involves maintenance action and is critical to the equipment maintaining its availability throughout the production period. This critical parameter identification should evolve through and be traceable to the PFMEA, process control plan, and the historical equipment and tool repair data.
3. Describe the corrective maintenance action necessary to control the critical design or process parameter for the piece of equipment involved in each operation of the manufacturing process.
4. Record the time interval or repair frequency necessary to maintain the equipment to its operational readiness relative to the identified maintenance action.
5. Provide a reaction or contingency plan to guide the appropriate personnel on the next course of action avoiding any catastrophic behavior to the equipment and subsequent downtime in the process.

Manufacturing Process: XYZ Printed Circuit Board Asm.
Supplier: ABC Company

Process Number	Machine, Device Jig, Tools for Manufacturing	Location of PM Equipment or Area	Description of Preventive Maintenance Procedure and Associated Tools	Preventive Maintenance Item					Reaction Plan
				Daily	Weekly	Monthly	Quarterly	Yearly	
5	Wave-soldering machine #2	Soldering bath	Remove residue that appears on solder of the bath with cleaning tool per instruction sheet #3	Once per 8-hour shift					Inform technician or manufacturing. Conduct cleaning procedure #2B prior to start-up
		Soldering bath —interior of machine	Remove dust and dirt per instruction sheet #4		Once per week on 1st shift				Same as above
		Reflector for heater	Remove the dust, dirt, and cloudiness with cleaner and cloth per instruction sheet #5			Each month at the beginning of 1st shift			Same as above
		Belt chain of soldering machine	Remove flux residue and solder residue			Each month at the beginning of 1st shift			Same as above
		Conveyor belt speed adjustment	Confirm the conveyor speed setting according to specification. Read value of meter mounted on machine.	Once every 2 hours					Stop machine and inform technician or manufacturing engineer.
		Solder temperature	Check meter reading of temperature per control plan	Once every 2 hours					Determine root cause, repair, and perform start-up per instruction #3C
		Preheat temp.	Validate temperature per set-up table for particular PCB	Once every 2 hours					Same as above
		Height of solder wave	Confirm height of solder wave per measurement scale apparatus #12 and instruction sheet #21	Once every 8 hrs.					Same as above

Figure 5.23 Scheduled Preventive Maintenance Plan—Printed Circuit Board Assembly Example

Benefits of the Tool

The benefits for utilization of preventive maintenance tools facilitates the assurance of tool and equipment operation. Some of the benefits for these tools are:

1. Tool and equipment studies provide assurance to the determination of the repair or maintenance interval for various equipment in the process.
2. Evaluating the time or cycle interval from previous repair and replacement data helps validate the scheduled PM plan.
3. Tool and equipment studies help to answer the questions regarding stability of a process operation by addressing the degree of variation over usage time as contributed by the maintainability of a tool or piece of equipment.
4. The scheduled preventive maintenance plan helps link important process parameter information transformed through the PFMEA and process control plan in order to assure process control and productivity of the assembly process throughout the production period.
5. The scheduled preventive maintenance becomes the main source of knowledge gained in the review of tool and equipment studies. It becomes the updated, central document at the technician and maintenance operation level for the assurance of tool and equipment stability.

H. Reliability Growth Management

What and Why the Tool Is Applied

Reliability growth management is a useful method for identifying failure modes, implementing design and process changes, and monitoring reliability progress on an ongoing basis during the early phases of a product development program. This tool involves: (1) the determination of reliability growth operating test environment or manufacturing assembly sequence, (2) execution and monitoring of test or build progress through the use of reliability growth models, (3) evaluation of test or build results and comparison to other analytical studies, (4) correction of design or process to eliminate defects found, and (5) confidence for the successful completion of verification and production build. The reason for the use of reliability growth management is to provide a mechanism to help build-in design and process reliability during the early stages of the product design and process development phase. By developing a reliability growth management process into the development phase, design and process reliability results can be continuously evaluated and monitored, improved, and benchmarked to other designs or processes thereby helping to increase the confidence on a successful product in the field.

When the Tool Is Applied

Reliability growth management is applied continuously and throughout the product design and process development phase of the PDP. The amount of time spent conducting reliability growth management depends on the point in which the product design and accompanying process meet the quality and reliability targets (that is, R90/C90, approximately 95%–100% FTC, etc.). This tool should help to increase the confidence of the product design and process to successively pass verification testing and process performance and capability.

Where the Tool Is Applied

Reliability growth management is applied to any design or process complexity level (component, subsystem, or system level). The implementation of the tool is most effective when it is applied to new or modified product designs and processes. Carryover products may not need to undergo such an effort since they have no new adaptations which may affect fit, form, or function.

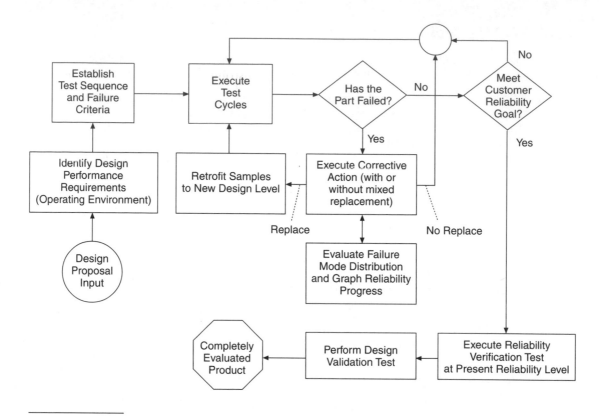

Source: John Bieda, "Reliability Growth Test Management in a Product Assurance Environment within the Automotive Component Industry," *1991 Proceedings—Annual Reliability and Maintainability Symposium* (New York, NY: © IEEE, 1991) p. 318. Adapted by permission.

Figure 5.24 Reliability Growth Test Management Process

Who Is Responsible for the Method

Responsibility for the application of reliability growth management is the product engineer, process or manufacturing engineer, and the product assurance engineer. The product and process engineers help set the test schedule, authorize the test or practice builds, determine the level of reliability growth, incorporate the improvements, etc. The product assurance engineer helps verify the test sequence and conditions, determines the growth curve, monitors its progress, performs analysis on the data obtained to compare against the growth curve, and helps out in the root cause effort for the product design and process.

How the Tool Is Implemented

The reliability growth management process can be implemented on product design and manufacturing process; however, the steps to be followed differ slightly for the product design and the process. The product design reliability growth process is described as follows (see flowchart in Figure 5.24):[11]

1. Identify appropriate design performance requirements, which would help to isolate the various environments necessary for test sequence development. Test environments should

be organized in a cyclical fashion so that test samples are exposed to many environments in a relatively short time.

2. Establish sample size and customer reliability goal levels. Assume or determine from field data the time-to-failure distribution. This failure distribution must be defined in order to calculate sample size. The determination of sample size should be a function of the demonstrated reliability goal and required customer confidence level:

$$N = -ln\frac{\left(1-CL\right)}{\left(1-R\right)} \qquad (25)$$

where: N = sample size
CL = required customer confidence level (%)
R = required reliability goal (%)

Assuming the exponential failure distribution:

$$R(t) = e^{-\lambda t} \qquad (26)$$

where: $R(t)$ = reliability as a function of time
λ = failure rate (failures/unit time)
t = operating period equivalent

3. Begin testing of samples per the reliability growth test environmental test sequence. Larger samples of later design levels should be put on test as available. Older parts should be left running if equipment space permits. Data of different design levels should be guided by the need to develop data on the latest design level parts.

4. Monitor and implement corrective action on failed parts during test execution.

5. Establish the reliability growth model (see example in Figure 5.25). A failure distribution from the test data should be determined and a growth slope calculated from initial failure information. An instantaneous MTBF or failure rate can be calculated once the growth slope is known. Determine the reliability growth slope in order to monitor the MTBF or reliability versus accumulated test time (see Figures 5.26a and 5.26b). For an exponentially distributed failure rate data case the instantaneous MTBF:

$$\theta(t) = \left(\frac{1}{k}\right)\left(\frac{t_a^b}{1-b}\right) \qquad (27)$$

where: $\theta(t)$ = instantaneous MTBF (unit time)
k = model coefficient (from method of least squares)
b = slope of the growth curve (from method of least squares)
t_a = accumulative test time (unit time)

If reliability is to be directly monitored versus accumulated test time, then an instantaneous failure rate calculation is needed:

$$\lambda_i = k(1-b)t_a^{-b} \qquad (28)$$

A best fit curve should be drawn as appropriate from this line equation.

6. Reliability or MTBF versus accumulated time should be plotted (confidence line also plotted) as failures occur and/or at regular time intervals. The first data point should be plotted at the first failure (see Figures 5.26a and 5.26b).

7. Growth testing should continue until the customer reliability goal is reached or an acceptable level is realized. Growth test charting should be updated as corrective actions occur.

Situation:

An electromechanical subsystem had undergone reliability growth testing under combined environmental test conditions. Nine samples were tested and exhibited failures during the test. After a specific test interval, all samples were removed, corrected, and placed back into the test sequence. After each corrective action phase, the total accumulative failures and test hours were used to determine the Duane reliability growth model parameters for the instantaneous mean time between failures (MTBF) or instantaneous failure rate. The method of least squares was specifically applied to obtain the slope (*b*), intercept (*a*), and model coefficient. Determine the MTBF and the reliability growth plot profile.

Solution:

Using the data from Table 5b the model parameters can be calculated based on the method of least squares. A summary of these results are as follows:

$$\text{slope } b = 0.9284$$
$$\text{intercept } a = 4.1607$$
$$\text{model coefficient } k = 64.1145$$
$$\text{accumulative test hours } t(a) = \text{various}$$
$$\text{mission time } t = \text{various}$$

The instantaneous mean time between failures (MTBF) model can be developed using the previously determined growth model parameters:

$$\theta_{i_a} = \left(\frac{1}{k}\right)\left(\frac{t_a^b}{1-b}\right)$$

$$\theta_{i_a} = \left(\frac{1}{64.1145}\right)\left(\frac{t_a^{0.9284}}{1-0.9284}\right)$$

Using the above equation the accumulative MTBF versus accumulative test time profile can be developed as was shown in Figure 5.26a. The instantaneous failure rates are necessary for determining the reliability as a function of accumulative test time. The following model was used to generate the instantaneous failure per accumulative test time:

$$\lambda_{i_a} = k(1-b)t_a^{-b}$$

For example, the above equation can be used to determine the instantaneous failure rate as a function of 17,000 accumulative test hours (see data in Table 5c):

$$\lambda_i(t_a = 17,000 \text{ accumulative test hours}) = 5.43 \times 10^{-4} \text{ failures/hour}$$

The reliability for each accumulative test hour can be calculated using the following model:

$$R_t = e^{-\lambda_{i_a}t}$$

Using the above equation the reliability versus accumulative test time profile can be developed as was shown in Figure 5.26b. For example, the above equation can be used to determine the reliability as a function of 552 hours of actual operating time:

$$R(t = 552 \text{ hours}) = 0.7410$$

Figure 5.25   Reliability Growth Test Management Example

TABLE 5b

Reliability Growth Model Parameters—Electromechanical Device Example

Functional Test Phase	Total Accum. Failure (r_a)	Total Accum. Test Hrs (t_a)	Growth Model Parametric Data			
			$Y =$ ln (r_a/t_a)	$X =$ ln (t_a)	X^2	$X \cdot Y$
I	94	180	−0.65008	5.19296	26.96683	−3.37584
II	111	3420	−3.42652	5.13740	66.21728	−27.88296
III	124	9306	−4.31999	9.13641	83.51054	−39.47784
IV	120	13869	−4.68530	9.53741	90.96219	−44.68563
	130	16965	−4.87174	9.73891	94.84637	−47.44544
SUM TOTAL			−17.95363	41.74509	362.50321	−162.86771

Notes: Growth slope parameters determined by the method of least squares:

Slope of growth slope curve $= -b = \dfrac{\Sigma(x \cdot y) - (\Sigma x \cdot \Sigma y)/N}{\Sigma x^2 - (\Sigma x)^2/N}$

Y intercept $= a = \frac{1}{N}[\Sigma y - b\Sigma x]$

A constant depending on equipment complexity (model coefficient) $= k = e^a$

TABLE 5c

Reliability Growth Test Plot Data—Electromechanical Device Example

Total Accum. Test Hours t_a	Instantaneous Failure Rate λ_i (f/Hr)	Reliability (95% Customer Usage Level)		
		$R(t = 552)$	$R(t = 2300)$	$R(t = 4600)$
180	.036975	1.4×10^{-9}	1.16×10^{-37}	1.36×10^{-74}
3420	.002403	.2654	.0040	1.58×10^{-5}
9306	.000949	.5922	.1127	.0127
13,869	.000655	.6966	.2217	.0491
16,965	.000543	.7410	.2868	.0823

Notes: Test time $t \neq t_a$:

    t_a = accumulated test hours for all samples together

    t = 95% customer usage operating period

Process reliability growth management involves monitoring process performance and capability indices or probability of success per operation for the total operation over time. A description of the technique is as follows (see example in Figure 5.27):

1. Decide on the sample size for the number of units to be produced during the assembly line process. Use the binomial approximation model to determine the number of samples that would be needed.

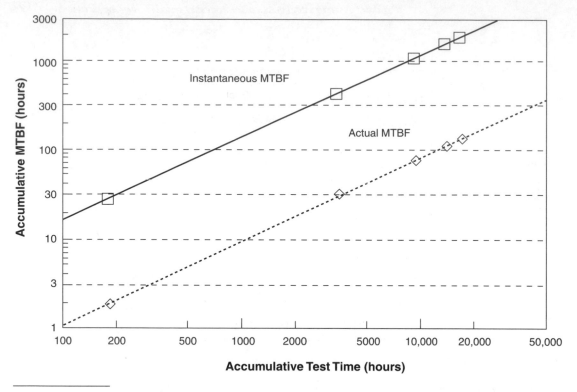

Source: John Bieda, "Reliability Growth Test Management in a Product Assurance Environment within the Automotive Component Industry," *1991 Proceedings—Annual Reliability and Maintainability Symposium,* (New York, NY: © IEEE, 1991) p. 319. Adapted by permission.

Figure 5.26a Reliability Growth Test Management—MTBF Distribution Profile

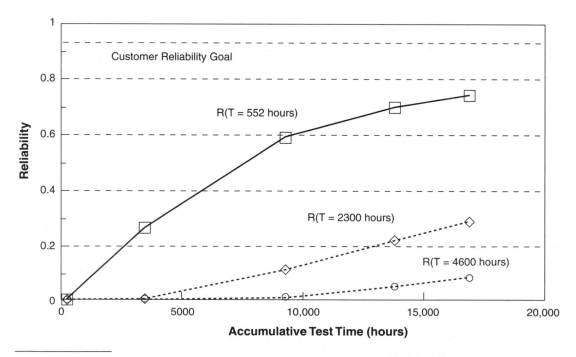

Source: John Bieda, "Reliability Growth Test Management in a Product Assurance Environment within the Automotive Component Industry," *1991 Proceedings—Annual Reliability and Maintainability Symposium,* (New York, NY: © IEEE, 1991) p. 319. Adapted by permission.

Figure 5.26b Reliability Growth Test Management—Reliability Profile Example

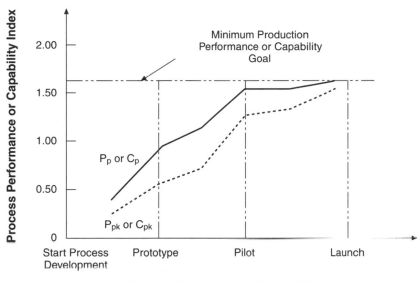

Figure 5.27 Process Reliability Growth Management Example

2. Determine the data type (attribute or variable) to be collected at each process station or for the total process. Refer to the process control plan to obtain the key process parameter to be measured and monitored (temperature, pressure, etc.). If the process characteristic is an attribute parameter, such as visually inspected parts per designated acceptance criteria, collect the number of acceptable parts per the total for the process.
3. Analyze the variable type data in terms of process performance or capability indices for each process station per process run. Analyze the attribute type data in terms of the probability of success for each process station per process run.
4. Make corrections to the process station or equipment based on the frequency of poor build. Record the next interval of data per the corrective action performed.
5. Plot the process performance or capability (or probability of success) as a function of the corrected process.
6. Make adjustments to the process in order to obtain the process reliability (process performance/capability or probability of success) objective in the established development period.

Benefits of the Tool

The benefits from the use of reliability growth management are significant. They involve the ability to improve product and process performance in a organized and productive fashion. Highlights of these benefit are:

1. Reliability growth management provides an insight into the accuracy of failure modes predicted, proportion of failure modes associated with each mode, reliability estimates from prediction, and verification of risks in a FMEA and FTA.
2. Reliability growth management provides a practice period early in the development cycle for evaluating designs and making improvements on these designs or processes before product design or process verification testing.
3. Reliability growth management helps to establish a base operating test environment from which accelerated tests may be developed in order to shorten the test sequence.
4. Reliability growth management allows the possibility for mature designs to receive verification in the latter stages or iterations of growth testing or process builds.

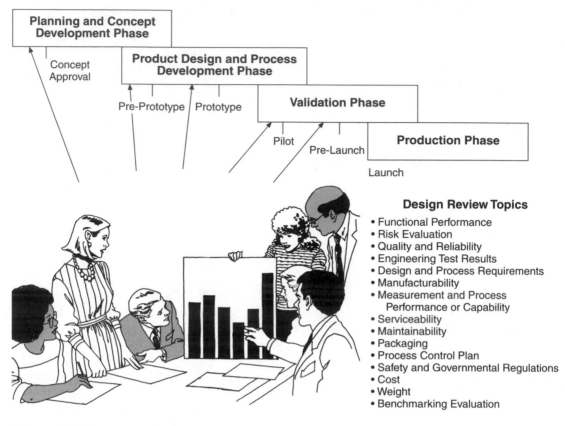

Figure 5.28   Design Reviews and Their Relationship to Product Development

I. Design Review

What and Why the Tool Is Applied

The fundamental objective of a design review is for early detection and remedy of design deficiencies which could jeopardize successful performance, quality and reliability, and manufacturability. The design review is an opportunity for the team to review the product design and process and evaluate the risks, improvements, and status of the product during development. It is important that the design review provide a forum for the exchange of test results, field information, benchmarking, etc., so that the team can evaluate and provide input to the risks associated with the design and process.

When the Tool Is Applied

Design reviews or risk assessments should be conducted several times throughout each phase of the product development process. The specific quantity of design reviews for a product are dependent on product complexity and criticality. The number of reviews can range from a minimum of 4–6 reviews per product development phase for simple products, to many reviews (8–12) for complex systems. The precise timing of a review is guided by the engineering milestone dates for product development. These reviews would be conducted at prescribed intervals prior to the actual product development milestone event (see Figure 5.28).

Where the Tool Is Applied

Design reviews are used on all product and process levels (new, modified, and carryover) and complexities (system, subsystem, and component).

Who Is Responsible for the Method

The membership can vary depending on the complexity of the product and size of the company. These members should include: designer, product releasing engineer, manufacturing engineer, procurement representative, material engineer, quality and reliability engineer (or product assurance specialist), field engineer, customer representative, etc. In large companies the chairman is generally the project manager or design engineer. In small companies it can be the head of the technical department.

How the Tool Is Implemented

Design reviews should be conducted in a informal setting and should involve some basic guidelines in order to make the meeting most effective. Some of these guidelines are listed in the following text:

1. Design review should not be viewed as an attack on the design or design organization. Fear should not be present. Deming had said in his 14 Points that fear should be driven out of the organization in order to promote trust and sharing of ideas.
2. Design reviews should be scheduled to allow time to react to the issues that are raised. A complete closure of the tasks and assignments given in the design review should be evaluated in a subsequent review.
3. Material to be discussed in the design review should be made available to the review team prior to the meeting.
4. Focus should be on the new applications or features that hold the *highest risk*. Test results during product design and process development and engineering studies such as benchmarking, simulation, or engineering development should be summarized and reviewed in the meeting. All FMEAs and fault tree analyses should reflect any activity performed to highlight new risks or reduce old risks.
5. Reviews should be formal and include meeting minutes. Assignments and deadlines should be reasonable and clear.
6. Detailed follow-up and closure of risks and opportunities must be provided. Management should be involved and kept responsible for any deficiencies in meeting these deadlines.
7. An agenda must be covered in each design review.
8. Several important questions should be covered in each design review. A sample of the questions to ask are:
   A. What are the internal and external customer requirements that the design will fulfill?
   B. Have customer requirements been thoroughly translated into engineering and production requirements?
   C. Will the design be similar to existing designs?
   D. What problems have been experienced with existing designs?
   E. What are likely design and process failure modes and what has been done to reduce risks?
   F. What are the greatest areas of risk?
   G. What is the expected level of reliability? Has it been achieved?
   H. Have critical manufacturing/assembly process characteristics been identified?
   I. How will processes be controlled?
   J. Is measuring equipment capable of monitoring processes?

K. What is the manufacturing process performance or capability of the total process? What is the process performance or capability of the individual assembly of manufacturing operations?

L. What has resulted from development and verification testing? Any new risks?

Shortcomings during a design review must be observed and avoided if the review is to be productive. Some of these shortcomings involve:

1. Designers resistance of non-peers participating in reviews (cultural pattern of design engineering groups).
2. Incomplete information: drawings, FMEA, test results, etc.
3. Lack of time to make meaningful, suggested changes.
4. "Not invented here" syndrome if reviews are mandated.

Benefits of the Tool

1. Design reviews provide a means to understand the progress and continuous improvement made on a product prior to the specific engineering milestone in a product development phase.
2. Design reviews facilitate the involvement of other engineering support to the development and improvement of the product and process.
3. Design reviews encourage awareness, communication, and teamwork between the product design and process engineering groups as well as other disciplines such as procurement and supply, marketing, etc.
4. Design reviews help to organize and structure the group toward addressing objectives and completing tasks.
5. Design reviews allow management to review the full development of the product and process and address issues with representatives of the organization all in one meeting.

NOTES

1. Chrysler, Ford, and General Motors Supplier Quality Requirements Task Force, *Potential Failure Mode and Effects Analysis (FMEA)—Reference Manual*, 1993.
2. Idaho National Engineering Laboratory, "Integrated Reliability and Risk Analysis (IRRAS), Version 1.0," June 1987.
3. Division of Risk Analysis, Office of Nuclear Regulatory Commission, *Measures of Risk Importance and Their Applications*, (Washington D.C., July 1983), pp. 1–8.
4. Boothroyd, G., and P. Dewhurst, *Product Design for Assembly*, (Wakefield, RI: Boothroyd and Dewhurst, 1987).
5. Chrysler, Ford, and General Motors Supplier Quality Requirements Task Force, *Advanced Product Quality Planning and Control Plan—Reference Manual*, 1994, pp. 33–56.
6. Chrysler, Ford, and General Motors Supplier Quality Requirements Task Force, *Measurement Systems Analysis—Reference Manual*, 1990, pp. 37–38.
7. Chrysler, Ford, and General Motors Supplier Quality Requirements Task Force, *Measurements Systems Analysis Reference Manual*, 1990, pp. 38–51.
8. Duncan, A. J., *Quality Control and Industrial Statistics*, 4th ed., Richard D. Irwin, Homewood, IL, 1974.
9. Chrysler, Ford, and General Motors Supplier Quality Requirements Task Force, *Measurement Systems Analysis—Reference Manual*, 1990, p. 79.
10. Kececioglu, Dimitri, PE, *Advanced Reliability Engineering—Maintainability Engineering*, 1986, Chapter 6, pp. 15–18.
11. Bieda, John, "Reliability Growth Test Management in a Product Assurance Environment within the Automotive Component Industry," 1991 Proceedings—Annual Reliability and Maintainability Symposium, (New York, NY: IEEE, 1991), pp. 317–321.

6

Product Assurance Tools and Processes and Their Application

Design/Process Validation Phase Tools

"When you can measure what you are speaking about, and express it in numbers, you know something about it; but when you cannot measure it, when you cannot express it in numbers, your knowledge is of meager and unsatisfactory kind; it may be the beginning of knowledge, but you have scarcely, in your thoughts, advanced to the stage of science."

> WILLIAM THOMSON, LORD KELVIN,
> "POPULAR LECTURES AND ADDRESSES," 1891–1894,
> JOHN BARTLETT, *Familiar Quotations,*
> LITTLE, BROWN AND COMPANY, 1982, P. 594.

Chapter Introduction

This chapter is the third of four chapters involving the strategic use of various quality and reliability tools during each of the product development phases. The purpose of this particular chapter is to discuss the practical application of some quality and reliability engineering techniques used to specifically evaluate test results obtained during the product design/process validation phase of the product development process (refer to the PAP worksheet example, Figure 3.3). The techniques covered in this chapter involve: (1) test data analysis, (2) reliability test data evaluation, (3) quality engineering evaluation, and (4) process review. In addition to the different kinds of test data evaluation, a discussion covering the elements involved in a manufacturing process review will be included to emphasize the importance behind a thorough evaluation of the manufacturing operation. Utilization of these techniques during the validation phase helps to verify the pilot assembly functionality of the design and stability of the process prior to production.

A. Test Data Analysis

1. Variable Type Data

What and Why the Tool Is Applied

The purpose of variable test data analysis is to obtain a better understanding of the trends and behavior of measurable data collected during testing. This measurable data may involve performance characteristics of a part which are affected by environmental stresses or natural variation. The approach used for the analysis of variable type data will primarily involve the evaluation and comparison of performance parameters collected before and after some test stimulus. One technique used to accomplish the comparison of variable data is the use of the test for statistical significance between the descriptive statistics gathered before and after measurable data. Another technique used in variable type data analysis is a visual comparison of the process performance statistics gathered from the before-and-after test data. The intent of both approaches is to quickly and simply determine if there may be a concern in the part performance as a result of the test condition applied. A significant difference or substantial shift in the process performance indice of the before-and-after test data may indicate a possible change in the part functionality.

When the Tool Is Applied

Variable test data analysis is applied during the development and validation phases of the product development process. Basically, these methods can be used to evaluate any variable type data collected throughout the PDP in order to help address questions regarding statistical significance of measurable data. Since the method is universal to any application, it may be used by nontechnical personnel in the organization to address questions regarding the behavior of certain variable type data.

Where the Tool Is Applied

Variable type data evaluation can be used to help analyze: 1) measurable data collected from component or system hardware tests and experiments, and 2) measurable data obtained from investigative studies in engineering, business, and marketing. These data analysis techniques are even applicable to variable data collected from software related products that control hardware functions.

Who Is Responsible for the Method

Variable test data evaluation would be conducted by any engineer, specialist, or technician working in a technical discipline or by any business or finance analyst working in nontechnical disciplines such as purchasing, marketing, etc. The results of such data analyses should be properly communicated to the appropriate personnel so that the best decisions can be made on product performance or marketability.

How the Tool Is Implemented

The execution of the variable data analysis can be divided into two sections: test for statistical difference in the mean and variance, or significant change in the process performance statistics of the before-and-after data distributions. This information can be summarized in a worksheet as shown in the example in Figure 6.1. An important consideration prior to analysis is that the data follow a normal distribution. Two tests for normality (goodness-of-fit) may be performed to help assess this condition. The Kolmogorov-Smirnov test is performed for sample sizes less than 25, and the chi-square test is used to help evaluate goodness-of-fit for sample sizes of 25 or more. Both methods should be used to validate the nature of the data distribution prior to test of significance or process performance evaluation.

Test Type and Description (include Test Standard #)	Test Stress Condition	Test Characteristic	Acceptance Criteria (nominal +/- specification limits)	Test Sample Description (before versus after stress)	Test Result (number of parts passed versus failed)	Descriptive Statistics		Hypothesis Testing for Significant Difference between the Means and Variances				Process Performance Statistics			Failure Mode Description	Failure Mechanism (root cause)
						Mean (X-bar)	Standard Deviation (s)	Type of Test	Calculated Test Value	Table (Critical) Value	Significant Difference (yes/no)?	Process Performance (P_p)	(P_{pk})	On-Target ($P_p - P_{pk}$)		
Mechanical and Electrical Endurance with Temperature and Humidity Cycling (TS-XY)	Temperature, Humidity, and Mechanical Cycling	Operating Torque (oz-in)	19.96 oz-in (maximum)	Before	10 Passed 0 Failed	9.4	1.174	F	3.451	3.18	Yes	3				
				After	10 Passed 0 Failed	7.8	0.632	t					6.41			
	Temperature, Humidity, and Power On/Off Switching	Voltage Drop (V)	0.10 V (maximum)	Before	10 Passed 0 Failed	55.5	0.527	F	1.19	3.18	No		28.15			
				After	10 Passed 0 Failed	55.7	0.483	t	0.865	1.734	No		30.57			
High Temperature Soak Test	60–85°C High Temperature Exposure	Current Draw (A)	Minimum 4.0 A Nominal 5.0 A Maximum 6.0 A	Before	6 Passed 0 Failed	4.8	0.322	F	1.69	6.39	No	1.035	0.828	0.207	None	
				After	5 Passed 1 Failed	5.2	0.248	t	2.203	1.86	Yes	1.345	1.075	0.28	Open Circuit	Overstressed driver circuit resistor R11 at 45°C

Source: Chrysler Corporation, *Test Data Analysis—Second Edition*, 1995, p. 22A. Adapted by permission.

Figure 6.1 Variable Test Data Analysis and Root Cause Spreadsheet—Example

The steps involved in conducting a test data analysis for determining the statistical difference before and after a test condition are described in the following text:

1. Conduct a test for normality from the data distribution in order to validate the interpretation of the results obtained for the test of statistical significance or process performance (see section C on process performance and capability studies for details).

   If the sample size is equal to or exceeds 25 pieces, then use the chi-square test technique (see example in Figures 6.2a and 6.2b):

   a. Develop hypotheses.
   b. Plot frequency histogram.
   c. Identify the level of significance (α).
   d. Determine the critical value of chi-square for the level of significance and degrees of freedom using Table 6a.
   e. Calculate the chi-square test statistic.
   f. Compare the test statistic to the critical value. Make decision of a normal versus a non-normal distribution. If the test statistic does not exceed the critical value then the distribution is normal, otherwise it is non-normal.

   If the sample size is less than 25 pieces, then use the Kolmogorov-Smirnov (KS) test technique (see example in Figure 6.3)[1]:

   a. Arrange data in order of increasing value.
   b. Determine sample size.
   c. Calculate the observed probability of each successive event.
   d. Calculate the expected probability of each successive event.
   e. Calculate the absolute difference between the observed and expected probability of events.
   f. Determine the maximum absolute difference (D_{max}).
   g. Determine the critical value from Table 6b based on sample size and level of significance.
   h. Compare calculated maximum difference to critical value.
   i. Make decision for normal versus non-normal distribution:

      - If calculated maximum difference is less than the critical value, then the normal distribution is accepted.
      - If calculated maximum difference is greater than the critical value, then the normal distribution is rejected.

2. Censor any samples from the data before and after the test stress that had exhibited failure. The sample sizes before and after must be equal in order to conduct a comparison test of the means and variances. Those samples that exhibited failure should be evaluated to determine the failure mechanism and corrective action.

3. Calculate the mean of the before-and-after sample test normal distribution:

$$\bar{x} = \frac{\sum\limits_{x=i}^{n} x_i}{n} \tag{29}$$

where: $\bar{x}$ = mean of sample measurements
   x_i = individual measurement for samples
   n = quantity of measurements

Problem:

Machine cycle times for 34 sample periods were recorded for a particular piece of equipment. Determine whether this sample represents a normal distribution.

Solution:

Since the number of samples exceeds 25, the chi-square test technique was applied to determine whether the data represent a normal distribution. The following steps are detailed below:

  a. A hypothesis was developed for the normal distribution case:

  H_o: Data represent a normal distribution
  H_1: Data *do not* represent a normal distribution

  b. The data was recorded in Figure 6.2b. From these data a standardized frequency distribution plot was developed.
  c. The 5 percent significance level was chosen for this chi-square test.
  d. The critical value of chi-square for the level of significance and degrees of freedom (see Table 6a):

  $$d.f. = k - m - 1 = 4 - 2 - 1 = 1$$

  where: $d.f.$ = degrees of freedom
  k = number of frequency distribution classes
  m = number of parameters estimated (i.e., mean and standard deviation)

  critical value @ 0.05 significance level = 3.841

  e. The computed chi-square test statistic is as follows:

  $$\chi^2 = \sum_{i=1}^{k}\left[\frac{(f(o)-f(e))^2}{f(e)}\right]$$ (30)

  where: $f(o)$ = observed class frequency in your sample
  $f(e)$ = expected class frequency assuming H_o is true

  Using the data in Figure 6.2b the chi-square test statistic can be calculated:

  $$\chi^2 = \frac{(-1.4)^2}{5.4} + \frac{(3.4)^2}{11.6} + \frac{(-1.6)^2}{11.6} + \frac{(-0.4)^2}{5.4} = 1.604$$

  f. The computed test statistic was compared with the crtitical value in order to determine a decision from the hypothesis test of normality. Since the test statistic (1.604) does not exceed the critical value (3.841), the hypothesis of normality cannot be rejected (accept hypothesis H_o). In other words, the data distribution represents a normal distribution.

Figure 6.2a Chi-square Test for Normality Example

Machine Cycle Time Data Set

Time (in seconds)

121.6	120.6
122.6	122.2
121.6	121.8
122.0	123.4
123.0	123.2
121.8	121.8
123.4	119.4
122.2	124.0
125.0	122.0
122.2	121.6
122.4	122.2
124.0	121.2
120.4	122.4
121.8	122.0
120.0	122.0
121.2	122.4
122.0	122.2

```
.... OPTION 1: NORMAL GOODNESS OF FIT TEST ....

       f(o)     f(e)      z Scale and Histogram
       -----    ------    +---------------------------
         0       0.0      :
                            -3+
         1       0.7      : *
                            -2+
         3       4.6      : ***
                            -1+
        15      11.6      : ***************
                            0+
        10      11.6      : **********
                            1+
         4       4.6      : ****
                            2+
         1       0.7      : *
                            3+
         0       0.0      :
```

Chi-square Test for Normality

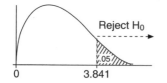

Reject H_0

.05

0   3.841

```
Standardized                                               2
Z-value       f(o)    f(e)    f(o)-f(e)   [f(o)-f(e)] /f(e)
----------    -----   ------  ---------   -----------------
Under -1.0      4      5.4      -1.4          0.360
-1.0 to 0.0    15     11.6       3.4          0.993
0.0 to 1.0     10     11.6      -1.6          0.222
1.0 and Over    5      5.4      -0.4          0.029

Total:         34     34.0       0.0          1.604

Chi-Square =    1.604  with d.f.=  1
```

Note: Details may not add to totals, due to rounding. In the test, the first 3 and last 3 classes were combined to enlarge f(e).

Source: David P. Doane, *User Manual for Exploring Statistics with the IBM PC,* (Addison-Wesley Publishing Company Inc., New York: 1985), pp. 109–110, 112. Reprinted by permission of Addison-Wesley Longman, Inc.

Figure 6.2b   Chi-square Test for Normality Example (Continued)

TABLE 6a

Abbreviated Table of Chi-Square Values

Quantity of Failures	Value of Chi-Square (χ^2)
0	4.605
1	7.779
2	10.645

This χ^2 table provides the lower 90% one-sided confidence limit values for 0, 1, or 2 failed test units. Consult a standard χ^2 table for a complete list of values.

Source: Chrysler Corporation, *Test Data Analysis*—Second Edition, 1995, p. 18.

Problem:

Two sets of machine cycle times (case #1– *time1* and case #2 – *time2*) for 15 sample periods were recorded for a particular piece of equipment. Determine whether each sample set represents a normal distribution.

Solution:

Since the number of samples does not exceed 25, the KS-test technique was chosen as a goodness-of-fit approach to determine whether each sample set represents a normal distribution. The computer software program "Statgraphics," version 4.0, was utilized to conduct the analysis. The following steps help describe the basic procedure and analysis of the data (see Figures 6.3b and 6.3c).

The application of the KS goodness-of-fit test was applied by selecting the distribution fitting analysis section of the Statgraphics computer program. A normal distribution was chosen as the type of data distribution being compared to each of the two sample data sets. The remaining steps are briefly described below:

a–d. The data of each case (*time1* and *time2*) were: (1) automatically arranged in increasing value for the sample size of 15 (steps a and b), (2) calculated for the observed probability of each successive event (step c), and (3) calculated for the expected probability of each successive event. These probabilities were based on the calculated mean and standard deviation as shown in the computer output.

e–f. The absolute difference between the observed and expected probability of events was automatically computed and a maximum absolute difference finally determined from these results. The maximum difference for *time1* was approximately 0.102 and for *time2* was 0.334.

Continued.

Figure 6.3a   Kolmogorov-Smirnov (KS) Test for Normality Example

g. The critical value from Table 6*b* based on sample size and a 0.10 level of significance was obtained. The critical value for *time1* and *time2* for $n = 15$ and a 0.10 level of significance was $D_{critical} = 0.304$.

h–i. A comparison was conducted between the calculated maximum absolute difference and the critical value for *time1* and *time2*. The following describes the decision for a normal versus a non-normal distribution:

time1 decision:

$D_{max} = 0.102 < D_{critical} = 0.304$ @ 0.10 level of significance. Therefore, we accept the null hypothesis (H_o) or agree that the sample period data set *time1 represents a normal distribution.*

time 2 decision:

$D_{max} = 0.334 > D_{critical} = 0.304$ @ 0.10 level of significance. Therefore, we reject the null hypothesis (H_o) or agree that the sample period data set *time2 does not represent a normal distribution.*

Machine Cycle Time Data Sets

Case 1	Case 2
time1	*time2*
2.4	1.5
5.6	2.0
5.4	9.4
4.5	2.3
2.5	2.7
6.2	2.1
6.7	9.0
8.4	9.6
5.8	1.9
3.2	9.4
4.7	2.4
6.8	8.8
9.1	1.9
7.4	9.3
7.2	2.3

Note: Computer software output used to conduct analysis was "Statgraphics," version 4.0. Statistical Graphics System, by Statistical Graphics Corporation, 1989.

Figure 6.3a *Continued.*

```
                          Distribution Fitting
------------------------------------------------------------------
Data vector: time1

Distributions available:
   (1)  Bernoulli           (7)  Beta           (13)  Lognormal
   (2)  Binomial            (8)  Chi-square     (14)  Normal
   (3)  Discrete uniform    (9)  Erlang         (15)  Student's t
   (4)  Geometric          (10)  Exponential    (16)  Triangular
   (5)  Negative binomial  (11)  F              (17)  Uniform
   (6)  Poisson            (12)  Gamma          (18)  Weibull

Distribution number: 14

Mean: 5.72667
Standard deviation: 2.00658



Results:

   Estimated KOLMOGOROV statistic DPLUS = 0.0960199
   Estimated KOLMOGOROV statistic DMINUS = 0.102002
   Estimated overall statistic DN = 0.102002
   Approximate significance level = 0.997659



Conclusion:
```

$D_{max} = 0.102 < D_{critical} = 0.304$ @ 0.10 level of significance. Therefore, we accept H_0 or agree that data represent a normal distribution.

Case 1:
Acceptance
of Normal
Distribution
Hypothesis

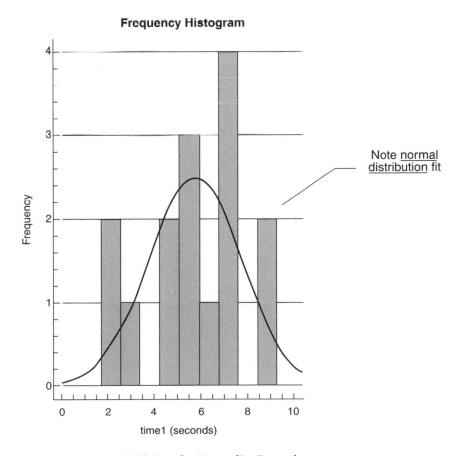

Figure 6.3b   Kolmogorov-Smirnov (KS) Test for Normality Example

```
                        Distribution Fitting
-----------------------------------------------------------------------
Data vector: time2

Distributions available:
   (1) Bernoulli          (7) Beta          (13) Lognormal
   (2) Binomial           (8) Chi-square    (14) Normal
   (3) Discrete uniform   (9) Erlang        (15) Student's t
   (4) Geometric         (10) Exponential   (16) Triangular
   (5) Negative binomial (11) F             (17) Uniform
   (6) Poisson           (12) Gamma         (18) Weibull

Distribution number: 14

Mean: 4.97333
Standard deviation: 3.62835
```

Case 2:
Rejection
of Normal
Distribution
Hypothesis

```
Results:

    Estimated KOLMOGOROV statistic DPLUS = 0.334523
    Estimated KOLMOGOROV statistic DMINUS = 0.25421
    Estimated overall statistic DN = 0.334523
    Approximate significance level = 0.0696644
```

Conclusion:

$D_{max} = 0.334 > D_{critical} = 0.304$ @ 0.10 level of significance. Therefore, we reject H_0 and agree that data do not represent a normal distribution.

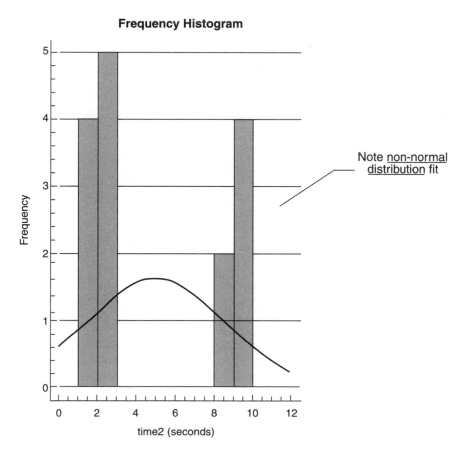

Figure 6.3c   Kolmogorov-Smirnov (KS) Test for Normality Example

TABLE 6b

Table of Critical Values of D in the Kolmogorov-Smirnov Goodness of Fit Test[2]

Sample Size	Level of Significance for $D = $ Maximum $\lvert F(x) - S_n(x) \rvert$				
	.20	.15	.10	.05	.01
1	.900	.925	.950	.975	.995
2	.684	.726	.776	.842	.929
3	.565	.597	.642	.708	.828
4	.494	.525	.564	.624	.733
5	.446	.474	.510	.565	.669
6	.410	.436	.470	.521	.618
7	.381	.405	.438	.486	.577
8	.358	.381	.411	.457	.543
9	.339	.360	.388	.432	.514
10	.322	.342	.368	.410	.490
11	.307	.326	.352	.391	.468
12	.295	.313	.338	.375	.450
13	.284	.302	.325	.361	.433
14	.274	.292	.314	.349	.418
15	.266	.283	.304	.338	.404
16	.258	.274	.295	.328	.392
17	.250	.266	.286	.318	.381
18	.244	.259	.278	.309	.371
19	.237	.252	.272	.301	.363
20	.231	.246	.264	.294	.356
25	.21	.22	.24	.27	.32
30	.19	.20	.22	.24	.29
35	.18	.19	.21	.23	.27
Over 35	$\dfrac{1.07}{\sqrt{n}}$	$\dfrac{1.14}{\sqrt{n}}$	$\dfrac{1.22}{\sqrt{n}}$	$\dfrac{1.36}{\sqrt{n}}$	$\dfrac{1.63}{\sqrt{n}}$

The values of D given in the table are critical values associated with selected values of n. Any value of D which is greater than or equal to the tabulated value is significant at the indicated level of significance.

4. Calculate the sample standard deviation of the before-and-after sample test distributions:

$$s \text{ or } \sigma_s = \sqrt{\frac{\sum\limits_{x=i}^{n}(x_i - \bar{x})^2}{(n-1)}} \tag{31}$$

where: s = sample standard deviation
x_i = individual measurement for samples
$\bar{x}$ = mean of the sample measurements
n = quantity of measurements

5. Perform the test for statistical significance by first testing for differences in the variances, F-test, and then testing for differences in the means, t-test (see Figure 6.4). The t-test can be conducted by treating the samples before and after the stress as independent samples of equal size ($n = n_1 = n_2$). The rationale for this treatment is that it is not truly known about the effect of the stress on the parts (stress may not be uniform for all parts). In addition, the treatment of the samples as being independent, as opposed to being paired, increases the sensitivity of the critical value when dealing with small sample sizes.

   The test of statistical significance for the difference in the variances between the two sample distributions is performed as follows:

   a. Calculate the F-test statistic by formulating the ratio of the larger and smaller sample variances:[3]

   $$F = \frac{S_{larger}^2}{S_{smaller}^2} \tag{32}$$

   where: F = F-test statistic
   s_{larger}^2 = largest of the before-or-after variance (s_{before}^2 and s_{after}^2).
   $s_{smaller}^2$ = smallest of the before-or-after variance (s_{before}^2 and s_{after}^2).
   s_{before} = standard deviation of the before-sample distribution
   s_{after} = standard deviation of the after-sample distribution

   b. Determine the F-critical value from table of F values (Table 6c) as a function of the number of degrees of freedom ($n - 1$). The sample size is the same before and after the stress is applied. Use an appropriate risk level, α, for a two-tailed test before referring to the degrees of freedom.

   c. Compare the calculated F-test statistic and the F-critical value. If the calculated F-test statistic is greater than the F-critical value, then there is a statistical significance between the variances of the before-and-after sample distributions. On the other hand, if the calculated F-test statistic is less than the F-critical value, then there is no statistical significance between the two sample distributions.

The test of statistical difference for the difference in the means between the two sample distributions is performed as follows:[4]

   a. Calculate the t-test statistic by first determining the standard deviation of the pooled measurements and then using this result to compute the t-test statistic. The pooling of the sample standard deviations is computed as follows:

   $$S_{pooled} = \sqrt{S_{before}^2 + S_{after}^2} \tag{33}$$

   where: s_{pooled} = standard deviation of the pooled measurements
   s_{before} = standard deviation of the pre-test measurements
   s_{after} = standard deviation of the post-test measurements

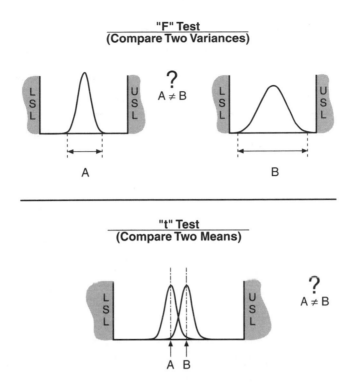

Figure 6.4   Tests of Significance

TABLE 6c

Abbreviated Table of *F* Critical Values for a Two-Tailed Test of Variance

Degrees of Freedom 1 $(n-1)$	*F* Critical Value	Degrees of Freedom 1 $(n-1)$	*F* Critical Value	Degrees of Freedom 1 $(n-1)$	*F* Critical Value
1	161.	9	3.18	17	2.28
2	19.00	10	2.98	18	2.22
3	9.28	11	2.82	19	2.17
4	6.39	12	2.69	20	2.12
5	5.05	13	2.58	21	2.09
6	4.28	14	2.48	44	1.66
7	3.79	15	2.40	229	1.00
8	3.44	16	2.34		

This table provides the *F* critical value for the respective number of degrees of freedom $(n-1)$. If the appropriate quantity does not appear, consult a standard F-Table for d/2 = 0.05 level of significance.

Source: Chrysler Corporation, *Test Data Analysis*—Second Edition, 1995, p. 12A.

TABLE 6d

Abbreviated Table of *t* Critical Values for a Two-Tailed Test of Means

Degrees of Freedom $(2n-2)$	*t* Critical Value	Degrees of Freedom $(2n-2)$	*t* Critical Value	Degrees of Freedom $(2n-2)$	*t* Critical Value
1	6.314	11	1.796	21	1.721
2	2.920	12	1.782	22	1.717
3	2.353	13	1.771	23	1.714
4	2.132	14	1.761	24	1.711
5	2.015	15	1.753	25	1.708
6	1.943	16	1.746	26	1.706
7	1.895	17	1.740	27	1.703
8	1.860	18	1.734	28	1.701
9	1.833	19	1.729	29	1.699
10	1.812	20	1.725	30+	1.697

This table provides the *t* critical value for the respective number of degrees of freedom $(2n-2)$. If the appropriate quantity does not appear, consult a standard T-table for d/2 = .05 level of significance.

Source: Chrysler Corporation, *Test Data Analysis*—Second Edition, 1995, p. 15A.

b. Calculate the t-test statistic by formulating the ratio between the difference of the sample means before and after testing and the standard deviation of the pooled measurements:

$$t = \frac{\bar{x}_{before} - \bar{x}_{after}}{\dfrac{S_{pooled}}{\sqrt{n}}} \qquad (34)$$

where: t = t-test statistic
S_{pooled} = standard deviation of the pooled measurements
$\bar{x}_{before}$ = mean of the sample measurements before the start of the test
$\bar{x}_{after}$ = mean of the sample measurements after the completion of the test
n = quantity of before-and-after test measurements ($n = n_{before} = n_{after}$)

c. Determine the t-critical value from a table of *t* values (Table 6d) as a function of the number of degrees of freedom ($n_{before} + n_{after} - 2$). Use an appropriate risk level, α, for a two-tailed test before referring to the degrees of freedom.

d. Compare the calculated t-test statistic and the t-critical value. If the calculated t-test statistic is greater than the t-critical value, then there is a statistical significance between the means of the before-and-after sample distributions. On the other hand, if the calculated t-test statistic is less than the t-critical value, then there is no statistical significance between the two sample distributions.

The steps involved in conducting test data analysis with process performance statistics can be divided into three areas: P_p, P_{pk}, and a measure of on-target, $P_p - P_{pk}$. These process performance indices are relative indicators of the relationship between the sample normal distribution and the specification limits. The process performance index P_p is an indicator of total variation or spread of the sample normal distribution as compared to the specification limits. The performance index P_{pk} is an indicator of location of the sample normal distribution as compared to the nearest specification limit. The difference between the process performance indices, $P_p - P_{pk}$, is an indicator of the mean of the sample normal distribution being coincident with the nominal specification value.

The calculation of these process performance indices is based on the computation of the sample mean and standard deviation as shown earlier in this section (equations 29 and 31).[5] See chapter 6, section C, part 1 for details on the process performance indices and their calculations.

Benefits of the Tool

Variable test data analysis is a valuable tool for evaluating engineering test data or developmental data. It is very beneficial when the analysis is presented in a simple manner to help interpret the variation and trends of descriptive statistics. The use of variable analysis facilitates a better understanding of the variation of parametric data and whether that variation is statistically significant to the performance of the device being studied. The following list highlights the benefits of variable test data analysis:

1. Variable test data analysis allows for a more descriptive means and detailed examination of the trends and variation of a particular design or process parameter. Results from such an analysis help to correlate parametric behavior to possible performance conditions or failure mechanisms.
2. Variable test data analysis provides more opportunities to evaluate data and generate conclusions on the behavior of a design or process. These methods descriptively indicate the degree of possible risk by the behavior of the variation in the data. Understanding this risk is very important to the engineering team during design reviews and risk assessment sessions in the PDP.
3. Variable test data analysis allows for thorough statistical analysis.

A. Test Data Analysis

2. Attribute Type Data

What and Why the Tool Is Applied

The purpose of attribute test data analysis is to facilitate interpretation of go/no-go type data and the solution processes available in evaluating such data collected during testing or during process capability studies. This kind of data would contain information regarding the proportion of acceptable versus unacceptable responses for a particular performance characteristic. The type of approach used to evaluate attribute type data would be application of the binomial and/or hypergeometric distributions. Both probability distributions would enable a calculation of a probability for a particular response given the sample size. Applying these techniques on attribute data over a particular operating time allows for a measure of stability.

When the Tool Is Applied

Attribute test data evaluation, similar to variable type data analysis, is applied during the development and validation phases of the product development process. Whether the data comes from a technical or nontechnical department, these methods are universal such that they may be applied to any application.

Where the Tool Is Applied

Attribute type data evaluation can be implemented to help interpret go/no-go type data from any kind of development or verification test, manufacturing capability study, marketing survey, etc. It is very useful when variable type data is not available from a developmental activity. The use of attribute type data analysis is very accommodating to the evaluation of software test data, which is usually in the form of attribute type output responses produced from a given combination of inputs.

Who Is Responsible for the Method

Anyone on the engineering team is responsible for performing this kind of data analysis. The product assurance engineer or quality/reliability engineer would be very useful as a consultant in addressing these probabilistic questions for the engineering team. In addition, he or she would assist in planning the experiments or studies which might necessitate the need for attribute data type evaluation.

How the Tool Is Implemented

Attribute data analysis can be accomplished several ways. Three specific methods can be used to evaluate the probability of an event given two mutually exclusive outcomes, such as defective or not defective, correct or incorrect, present or absent, acceptable or not acceptable. The first method essentially involves calculating a probability of success or yield. The second method involves the use of the binomial distribution, and the third method involves the use of the hypergeometric distribution. The binomial distribution is appropriate for sampling with replacement from a large population.[6] The hypergeometric distribution is appropriate for sampling without replacement from a small population.[7] A description of the three methods is described as follows (see examples in Figure 6.5):

Method #1: Basic Proportion Technique

1. Establish the basic criteria of what constitutes an acceptable part versus an unacceptable part.
2. Count the number of successful or unsuccessful outcomes (event x) from the total sample size.
3. Calculate the probability of the particular event by forming a ratio of the total number of acceptable outcomes (event x), or particular event of interest, and the total number of samples:

$$P(x) = \frac{\text{Total Number of Outcomes for Event } X}{\text{Total Number of Samples}} \tag{35}$$

where: $P(x)$ = probability of event x
x = event x (successful outcomes or unsuccessful outcomes)

Method #2: Binomial Distribution Technique

1. Identify the requirements for *sampling with replacement* and determine the acceptance criteria for the samples randomly selected.
2. Assure that the following conditions are met before applying the binomial distribution model:
   a. Each trial results in one of two possible, mutually exclusive outcomes.
   b. The probability of success, p, remains constant from trial to trial. The probability of failure is $1 - q$.
   c. The trials are independent; that is, the probabilities associated with any particular trial are not affected by the outcome of any other trial.
3. Randomly sample n objects from the given population.

4. Determine the probability of interest using the binomial distribution model:

$$P(x) = \frac{n!}{x!(n-x)!} q^{n-x} p^x \tag{36}$$

where: $P(x)$ = probability of event x
x = event of interest
n = number of samples taken with replacement
p = probability of a success
q = probability of a failure

Method #1

A manufacturing process run was performed for a particular design. The total number of parts processed through station #4 was 450. Out of the 450 parts assembled at station #4 there were 24 parts that were rejected by the inspector who was assigned to check the electrical and mechanical parameters of the part after assembly at station #4. What was the probability of success at station #4?

Using equation #35, the probability of success is determined:

$$P(x) = \frac{\text{Total Number of Outcomes for Event } X}{\text{Total Number of Samples}}$$

$$P(x) = \frac{426 \text{ Acceptable Samples}}{450 \text{ Total Number of Samples}} = 94.7\%$$

Method #2A

A quality engineer has determined that the average yield of a particular process is 85% acceptable parts. Suppose that the engineer randomly selects 10 parts from the process. What is the probability that there will be two or fewer unacceptable parts from the sample?

Using equation #36, the probability of two or fewer parts from the sample can be determined:

$$P(x) = \frac{n!}{x!(n-x)!} q^{n-x} p^x$$

where: $P(x)$ = Probability of event X
x = Event of interest = two or fewer unacceptable parts
n = Number of samples taken with replacement = 10
p = Probability of failure = 1.00 − 0.85 = 15%
q = Probability of success = 85%

The variable p is denoted as the probability of failure for this particular problem. Thus, the variable q becomes the probability of success for this case.

Continued.

Figure 6.5   Attribute Data Analysis Examples

The probability of two or fewer failures can be mathematically interpreted as:

$$P(x \le 2) = P(x = 0) + P(x = 1) + P(x = 2)$$

Using equation #36 for $x = 0$, 1, and 2 failures:

$$P(x < 2) = \frac{10!}{0!(10-0)!}0.85^{10-0}0.15^0 + \frac{10!}{1!(10-1)!}0.85^{10-1}0.15^1 + \frac{10!}{2!(10-2)!}0.85^{10-2}0.15^2 = 82.02\%$$

Method #2B

A quality engineer has determined that the average yield of a particular process is 85% acceptable parts. Suppose that the same engineer randomly selects 12 parts from the process. What is the probability that there will be exactly two unacceptable parts?

Using the same equation #36, the probability of exactly two failures is determined:

$$P(x = 2) = \frac{12!}{2!(12-2)!}0.85^{12-2}0.15^2 = 0.2924 \text{ or } 29.24\%$$

Method #3

A package of six electronic modules contains two that are defective and four that are nondefective. A quality engineer selects three modules at random from the package. What is the probability that the sample contains exactly one defective module?

Using equation #37, the probability of exactly one defective is as follows:

$$P(x = 1) = \frac{\left(\frac{2!}{1!(2-1)!}\right)\left(\frac{4!}{(3-1)!(4-(3-1))!}\right)}{\left(\frac{6!}{3!(6-3)!}\right)} = 0.60 \text{ or } 60\%$$

Figure 6.5 *Continued.*

Method #3: Hypergeometric Distribution Technique

1. Identify the requirements for *sampling without replacement* and determine the acceptance criteria for the samples randomly selected.
2. Select at random a sample size n from the small, finite population.
3. Determine the size of the elements of one type N_1 (successful outcomes) and the elements of the other type N_2 (unsuccessful outcomes).
4. Establish the question for the probability of the number of successful outcomes x from the sample.
5. Apply the hypergeometric distribution model to calculate the probability of success:

$$P(X = x) = \frac{\left(\frac{N_1}{x!(N_1-x)!}\right)\left(\frac{N_2}{(n-x)!(N_2-(n-x))!}\right)}{\frac{N}{n!(N-n)!}} \tag{37}$$

where: $P(X = x)$ = probability that the sample will contain x successes
N_1 = number of elements of one type (successes) from population N
N_2 = number of elements of another type (failures) from population N
N = population of size N
n = sample size without replacement

Benefits of the Tool

An understanding of attribute data analysis is essential in addressing probabilistic questions regarding go/no-go type data. Even though variable type data is more informative from a statistical perspective than attribute type data, knowledge and proficiency in applying attribute data analysis methods is important. Due to the need to simplify data collection and measurement activities, it is necessary to interpret and consult on the results from studies involving attribute type data. The following list describes the benefits:

1. Attribute type data analysis provides a quantitative approach to assess information which is not as descriptive as variable type data.
2. Attribute type data analysis helps to focus the evaluator on the background information accompanying the data, such as the population size, replacement technique, etc.
3. Attribute type data analysis allows for very adaptable approaches to the solution of probabilistic inquiries from engineering or other technical organizations.
4. Attribute type data analysis provides a relatively simple calculation process for the determination of a probability.

B. Reliability Test Data Evaluation

1. Stress/Strength Interference

What and Why the Tool Is Applied

Stress-strength interference for reliability evaluation essentially involves an analysis of a given failure mode for a component by comparing the stress and strength probability distributions and identifying the unreliability. The stress distribution represents the applied environmental loading condition (mechanical force, fluid pressure, etc.), and the strength distribution represents the ability to withstand this loading effect. An unreliability estimate may be determined by identifying the intersection of the two probability distributions and calculating the unreliability from the resulting probability of failure area. This reliability analysis technique is most appropriate for mechanical components in which the characteristics of stress and strength are common design features that may be available or determined by measurement. The results of stress-strength interference analyses are valuable to the verification of design intent and development of comprehensive failure rate models that may help to predict the relationship of several design parameters to the component or assembly reliability.

When the Tool Is Applied

Stress-strength interference is used to evaluate products during the design/process development and validation phases of a product development process. This particular tool can be applied successively throughout these phases to help gauge the improvement in a design based on adjustments of particular design features and parameters. It is a very useful tool when it is necessary to examine the potential for failure based on the location and shape of the stress and strength frequency distributions during accelerated testing or when the failure point on a product is not physically obvious.

Where the Tool Is Applied

Stress-strength is applied to any process or process complexity level (component, subsystem, or system level) and part status (new, modified, or carryover). The use of stress-strength for carry-over products may be applied since the test or field environment may impress a unique stress distribution and therefore shift the stress and strength distributions and influence the determination of unreliability.

Who Is Responsible for the Method

The practitioner of such a tool would most likely be the product assurance specialist or quality/reliability engineer. His or her role is to help define and analyze the stress and strength distributions determined through testing. Cooperation from the supplier, design, product, and process engineering would be necessary in the communication of inputs and outputs from the study as well as for application of the results into the product design or process.

How the Tool Is Implemented

Conducting a stress-strength analysis of a product design involves determining the probability distributions of stress and strength and then evaluating the location of these two distributions for the understanding of the probability of failure. The parameters for the stress and strength distributions can be found in testing or by previous knowledge. Application of this form of analysis is described in the following text and is demonstrated through the use of a computer software program with output shown in the example shown in Figure 6.6:[8]

1. Perform parametric test to generate the stress/strength data distributions under field usage conditions, or obtain data from similar applications.
2. Identify the type of frequency distribution for the stress and strength (normal, lognormal, Weibull, etc.) and the descriptive statistics which help describe it, such as the mean and standard deviation.
3. Plot the frequency distributions for both the stress and strength data. Compare the two distributions and identify the area which may be intersected by the two distributions. This intersection represents the probability of failure. If the two distributions do not intersect, then there is no risk of failure for the product based on the data collected and the conditions surrounding this collection.
4. Calculate the reliability or unreliability of the product design based on the probability of failure area indicated by the intersection. A reliability of 100 percent would be achieved if the stress and strength distributions do not intersect.
5. Interpret the relationship between the stress and strength distributions and adjust the design or process to achieve a strength distribution which exceeds the stress distribution. The further the strength distribution exceeds the stress distribution the greater the safety margin and the higher the reliability. See Figure 6.7 for the various strategies in improving reliability based on stress-strength analysis.

Benefits of the Tool

Stress-strength analysis provides significant advantages in the ability to facilitate design improvement decisions during development and validation. It is a relatively simple method for graphically or analytically understanding the relationship of a stress and strength distributions. Some of the detailed benefits are as follows:

1. Stress-strength analysis allows for a simple, graphical view of the relationship between the stress and strength distributions. It enables the engineer to make a quick assessment of the reliability of the product design by indicating how much effort might be necessary to increase the strength characteristics for the part.

Stress (Normal), Strength (Normal) Distributions

Stress		Strength			
Mean	57.947	Mean	69.284	Reliability	0.98705
Std. Dev.	4.837	Std. Dev.	1.582	Unreliability	0.01295

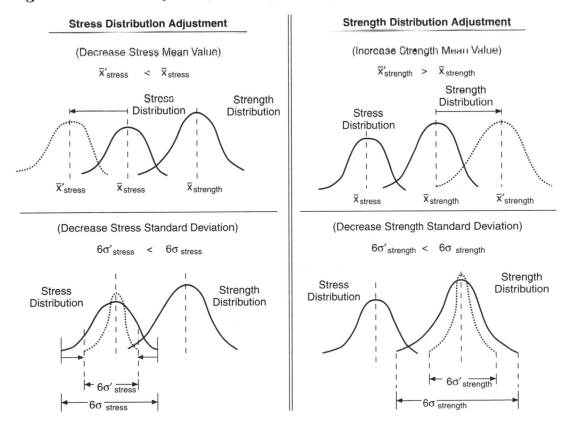

Source: John Bieda and Michael Holbrook, "Reliability Prediction, The Right Way", *Reliability Review,* a publication of the Reliability Division, ASQ, Vol. II, June 1991, p. 9.

Figure 6.6   Stress/Strength Analysis Technique—Example

Figure 6.7   Reliability Improvement through Stress/Strength Analysis

2. Stress-strength analysis provides some relative measure for the engineer on the safety margin between the mechanical strength characteristics and stress.
3. Stress-strength analysis generates a quick verification of a design's strength to a given stress condition.
4. Stress-strength analysis permits flexibility in the choice of frequency distributions in the analysis of the stress and strength distributions.
5. Stress-strength analysis facilitates the development of failure rate models and promotes the accumulation of a priori knowledge, which is helpful for the development of future designs.

B. Reliability Test Data Evaluation

2. Reliability Demonstration Test Analysis

What and Why the Tool Is Applied

Reliability demonstration test analysis requires the evaluation of life test data to determine the reliability of a design for comparison to a reliability requirement. This process specifically involves the use of statistical models and equations to help assess component, subsystem, or system performance as experienced during a life test. These analysis approaches must consider the number of failures exhibited during a given life test duration. The particular situations that may occur regarding the number of failures and the type of analysis to calculate a reliability are:

    A. *Tests with no failure:* Chi-square method
    B. *Tests with two or less failures:* Weibull analysis method
    C. *Tests with three or more failures:* Weibull graphical analysis method

When the Tool Is Applied

The reliability demonstration analysis techniques can be applied during the validation phase and in various instances throughout the design and process development phase in the PDP. The specific periods during the product design and process development phase in which these methods are applied can be during development testing. A reliability assessment may be made from the accelerated development test to assess design feasibility under given environmental conditions.

Where the Tool Is Applied

Reliability demonstration test analyses can be applied to any design complexity (component, subsystem, or system level) or part status (new, modified, or carryover).

Who Is Responsible for the Method

The principal parties for the application of such tools would involve the reliability, quality, or product assurance specialist. The PA specialist would assist the design, product, or process engineer in the review of the test data during development or verification testing and evaluate the reliability.

How the Tool Is Implemented

The application of either of the three analysis approaches is based on the number of failures:

A. Test with no failures: Chi-square method
    For the test with no failures, the chi-square method can be used to approximate the reliability. This method may be used to interpret results for a time-terminated, design life cycle, or extended life scenario. It is recommended that this approach be used for no failures only since a chi-square estimate for some number of failures may produce a too conservative estimate. The analytical technique is described in the following text (see example in Figure 6.8):
    1. Calculate the demonstrated failure rate:

Example #1—Tests With Zero Failures

An electronic circuit board assembly has a reliability demonstration target of $R = 90\%$, $C = 90\%$. One design life of testing is equivalent to 1000 hours of a multistress environmental test exposure. Twenty-two samples were tested to the multistress test sequence for 1000 hours. Five samples from the 22 were continued on test to three design lifes or 3000 hours without failures. What is the demonstrated reliability?

Using equation #38 and the χ^2 value for r = 0 failures (see Table 6a):

$$\lambda = \frac{\chi^2}{2T_a}$$

$$\lambda = \frac{4.605}{2(17 \times 1,000 + 5 \times 3,000)}$$

$$\lambda = 71.953 \times 10^{-6} \, Failures / Hour$$

The next step is to determine the demonstrated reliability from equation #39 using the failure rate previously calculated:

$$R = e^{-\lambda t}$$
$$R = e^{-(71.953 \times 10^{-6})(1,000)}$$
$$R = 0.93057 \text{ or } 93.05\%$$

The calculated reliability exceeds the reliability demonstration target of 90%.

Example #2—Tests With Two or Fewer Failures

2A. One Failure Case:

The same electronic circuit board assembly as in example #1 is to be tested to the reliability demonstration target of $R = 90\%$, $C = 90\%$. All 22 samples passed one design life or 1000 hours of the multistress environmental test exposure. Five samples were continued on test and produced one failure at 1800 hours. The remaining four samples continued through three design lifes (3000 hours) without failure. What is the demonstrated reliability?

The first step is to calculate the maximum likelihood estimate of the Weibull scale parameter using equation #40. A Weibull slope (β) = 3.0 was chosen arbitrarily to represent an increasing failure rate behavior later in test:

$$\bar{\alpha} = \left(\sum_{i=1}^{n} \frac{T_i^\beta}{r} \right)^{\frac{1}{\beta}}$$

$$\bar{\alpha} = \left(\frac{17 \times 1000^3 + 1 \times 1800^3 + 4 \times 3000^3}{1} \right)^{1/3}$$

$$\bar{\alpha} = 5076.581$$

The next step is to calculate the 90% lower confidence limit for the scale parameter using equation #41 and substituting into this equation the previous result for α. The χ^2 value for $r = 1$ failure is 7.779 (see Table 6a):

$$\alpha_{90} = \bar{\alpha} \left(\frac{2r}{\chi^2} \right)^{1/\beta}$$

$$\alpha_{90} = 5076.581 \left(\frac{2 \times 1}{7.779} \right)^{1/3}$$

$$\alpha_{90} = 3228.049$$

Continued.

Figure 6.8 Reliability Test Analysis Examples

The last step is to calculate the demonstrated reliability for one design life using equation #42 and the previous result for α_{90}:

$$R = e^{-\left(\frac{t}{\alpha_{90}}\right)^{\beta}}$$
$$R = e^{-\left(\frac{1000}{3228.049}\right)^{3}}$$
$$R = 0.9707 \text{ or } 97.07\%$$

The calculated reliability exceeds the reliability demonstration target of 90%.

2B. Two Failure Case:

The same electronic circuit board assembly as in example #1 is to be tested to the reliability demonstration target of $R = 90\%$, $C = 90\%$. All 22 samples passed one design life or 1000 hours of the multistress environmental test exposure. Five samples were continued on test and produced a failure at 1700 hours and 2400 hours. The remaining three samples continued through three design lifes (3000 hours) without failure. What is the demonstrated reliability?

The first step is to calculate the maximum likelihood estimate of the Weibull scale parameter using equation #40. A Weibull slope (β) = 3.0 was chosen arbitrarily to represent an increasing failure rate behavior later in test:

$$\bar{\alpha} = \left(\sum_{i=1}^{n} \frac{T_i^{\beta}}{r}\right)^{\frac{1}{\beta}}$$
$$\bar{\alpha} = \left(\frac{17 \times 1000^3 + 1 \times 1700^3 + 1 \times 2400^3 + 3 \times 3000^3}{2}\right)^{1/3}$$
$$\bar{\alpha} = 3879.057$$

The next step is to calculate the 90% lower confidence limit for the scale parameter using equation #41 and substituting into this equation the previous result for $\bar{\alpha}$. The χ^2 value for $r = 2$ failures is 10.645 (see Table 6a):

$$\alpha_{90} = \bar{\alpha}\left(\frac{2r}{\chi^2}\right)^{1/\beta}$$
$$\alpha_{90} = 3879.057\left(\frac{2 \times 2}{10.645}\right)^{1/3}$$
$$\alpha_{90} = 2799.181$$

The last step is to calculate the demonstrated reliability for one design life using equation #42 and the previous result for α_{90}:

$$R = e^{-\left(\frac{t}{\alpha_{90}}\right)^{\beta}}$$
$$R = e^{-\left(\frac{1000}{2799.181}\right)^{3}}$$
$$R = 0.9554 \text{ or } 95.54\%$$

The calculated reliability exceeds the reliability demonstration target of 90%.

Figure 6.8 *Continued.*

$$\lambda = \frac{\chi^2}{2T_a} \tag{38}$$

where: λ = demonstrated maximum failure rate (units/time)
χ^2 = chi-square value from Table 6a
T_a = accumulated test unit-duration (cycles, hours, etc.)

2. Calculate demonstrated reliability using the previously determined failure rate:

$$R(t) = e^{-\lambda t} \tag{39}$$

where: $R(t)$ = demonstrated maximum reliability
λ = demonstrated maximum failure rate (units/time)
t = length of specified mission life (cycles, hours, etc.)
e = base of natural logarithms equal to 2.7182

B. Tests with two or less failures: Weibull analysis method

For the test with two or less failures, the Weibull analysis method can be used to approximate the reliability. This method may also be used to interpret results for a time-terminated, design life cycle, or extended life scenario. The maximum likelihood estimate for the scale parameter or characteristic life will be used to focus on the later failure incidents (see example in Figure 6.8):

1. Determine the Weibull scale parameter[9]:

$$\overline{\alpha} = \left[\sum_{i=1}^{n} \frac{T_i^\beta}{r} \right]^{\frac{1}{\beta}} \tag{40}$$

where: $\overline{\alpha}$ = Weibull scale parameter estimate
T_i = test time for the ith sample
r = number of failures during test
β = Weibull slope (assumed or predetermined)

2. Calculate the percent confidence limit for the previously determined scale parameter:

$$\alpha_{x\%} = \overline{\alpha} \left(\frac{2r}{\chi^2} \right)^{\frac{1}{\beta}} \tag{41}$$

where: $\alpha_{x\%}$ = lower or upper percent confidence limit of the scale parameter
$\overline{\alpha}$ = Weibull scale parameter estimate
r = number of failures during test
β = Weibull slope (assumed or predetermined)
χ^2 = chi-square value from Table 6a

3. Calculate the demonstrated reliability:

$$R(t) = e^{-\left(\frac{t}{\alpha_{x\%}} \right)^\beta} \tag{42}$$

where: $R(t)$ = demonstrated reliability as a function of mission life
$\alpha_{x\%}$ = lower or upper percent confidence limit of scale parameter
t = length of specified mission life (cycles, hours, etc.)

β = Weibull slope (assumed or predetermined)
χ^2 = chi-square value from Table 6a
e = base of natural logarithms equal to 2.7182

C. Test with three or more failures: Weibull graphical analysis method (see chapter 6, section B, part 3).

Benefits of the Tool

The primary benefit of these reliability demonstration analysis tools is to facilitate the evaluation of those tests performed in the validation phase of the PDP. These analytical tools may also be applied to address reliability questions from development testing during the product design and process development phase. The following text lists some detailed benefits for the utilization of these tools:

1. Reliability demonstration test analysis provides the opportunity to calculate reliability based on a specific number of failures.
2. Reliability demonstration test analysis provides useful models for evaluating reliability as a function of failure distribution parameters and the mission period.
3. Reliability demonstration test analysis enables quick application of the model to a given set of data gathered during product life testing.
4. Reliability demonstration test analysis includes confidence limits and estimators to the parameters of the distribution when limited failures are experienced.

B. Reliability Test Data Evaluation

3. Weibull Distribution Analysis Technique

What and Why the Tool Is Applied

The Weibull distribution analysis technique is a very useful and versatile statistical tool for determining specific product life characteristics of various data distributions and allowing for the calculation of failure rate and reliability during testing. It is the third technique as discussed in the previous section for determining reliability. This form of analysis involves the identification of the product life parameters for any given distribution through a graphical approach. Weibull analysis is very useful in describing the relationship between the number of failures or failure rate and the characteristics found in the reliability bathtub shaped curve (see Figure 6.9). The Weibull distribution parameters are used in either a failure rate or reliability equation to further describe the life characteristics of the product design during some given environmental stimulus (field or test). The reliability result from such an analysis provides the basis for comparing design alternatives or improving previous design configurations by helping to evaluate the life parameters for a given product or process.

When the Tool is Applied

The Weibull distribution technique can be applied throughout design and process development or during the validation phase of the product development process. It is a tool that can be applied to any time or cycles to failure data to provide insight into the behavior of a product under some given stress.

Where the Tool Is Applied

Weibull analysis can be applied to any design or process complexity level (component, subsystem, or system level) and part status (new, modified, or carryover).

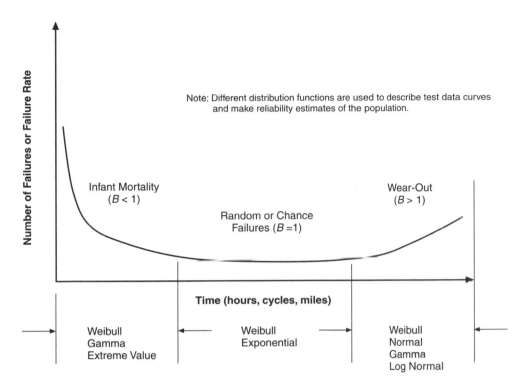

Figure 6.9 Reliability Bathtub Curve and Its Implications

Who Is Responsible for the Method

Principal parties for the application of such a tool would involve the reliability, quality, or product assurance specialist. These individuals would assist the design, product, or process engineer in reviewing the test data during development or verification testing and evaluate the failure mode distributions and resulting reliability. The quality engineer may apply the Weibull distribution to evaluate process or quality test data.

How the Tool Is Implemented

The Weibull distribution analysis technique can be performed using a graphical estimation approach. This technique will allow the determination of key characteristics from a variable type data distribution. If the distribution represents a time to failure behavior, then the resulting failure mode distribution characteristics can be used to identify the reliability. The steps for performing a two-parameter (minimum life equals zero) graphical analysis from time or cycle to failure data are outlined in the following text (an example of a Weibull analysis utilizing a computer software program "Weibull Smith" by Fulton Findings can be referred to in Figure 6.10).[10]

1. Gather the time or cycle to failure data from the life test and decide on how the data are to be grouped prior to performing the analysis. The data may be grouped by similar failure modes or other conditions. It should be noted that the results of the analysis will reflect only those type of data grouped and may not reflect other behaviors of the parts under examination. The analyst may choose to evaluate all failures on one plot or subdivide the analysis by individual plots based on each unique failure mode condition.

Problem:

An electromechanical switch assembly had undergone an accelerated life test for design validation. Five samples were tested to first failure under a combined powered-temperature and vibration environmental stress condition (see Figure 6.11a). What is the inherent reliability of the design for a 10 year operating duration?

Solution:

The five data points for switch cycles to first failure were recorded ($n = 5$, suspended $s = 0$), and an analysis was conducted using Weibull Smith software. The median rank values for each failure were determined and a Weibull plot was generated. The results for the Weibull slope and characteristic life were obtained (see Figure 6.11b). A correlation coefficient value (r^2) of 91.2% was obtained for the regression fit line and double-sided 90% confidence limits. Based on these results a reliability was calculated as follows:

Using the Weibull distribution two-parameter equation #43 (minimum life = 0):

$$R(t) = e^{-\left(\frac{t}{\theta}\right)^{\beta}}$$

and substituting the Weibull distribution parameters as determined from the Weibull analysis:

        t = 10 year (95th percentile usage level) equivalent = 4,518,360 cycles
        β = Weibull slope (shape parameter) = 3.03
        θ = Characteristic life = 9,069,967 cycles

The switch reliability can be calculated:

$$R(t = 4{,}518{,}360) = e^{-\left(\frac{4{,}518{,}360}{9{,}069{,}967}\right)^{3.03}} = 88.60\%$$

Figure 6.10   Weibull Distribution Analysis Example

2. Organize and rank the failure data points from the lowest to highest numerical value.
3. Use the a table of median ranks to assign a particular median rank value to each of the ordered failure data.
4. Either plot on Weibull probability paper (may include Weibull slope scale) each of corresponding median rank values from the original data or use the computer program "Weibull Smith" to enter failure data directly and obtain a Weibull plot (see Figure 6.11).
5. Draw a straight line through the plotted points to manually approximate the trend of the data. This line should be the "best fit" line through each of the points. Two other more statistical methods of fitting a line through the plotted points involve a regression or maximum likelihood approach. The regression line approach favors those failures which occurred early, whereas the maximum likelihood approach favors those failures which occurred later in the test. The Weibull Smith software program automatically allows for both the of these curve-fitting approaches.
6. Draw a line parallel to the median ranks vertically fitted line and through the Weibull slope scale for the manual approximation approach. Read the Weibull slope or distribution shape parameter β ("Weibull Smith" automatically determines the slope). This

Cycles to Failure Data and Median Rank Values

Cycles to Failure	Median Rank Value
4,924,800	0.1296
5,637,600	0.3148
9,201,600	0.5000
9,590,400	0.6852
11,145,600	0.8704

Figure 6.11a   Weibull Graph Data—Electromechanical Switch Assembly Life Test Example

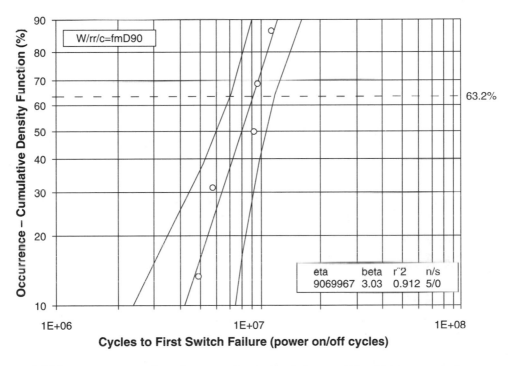

Figure 6.11b   Weibull Graph—Electromechanical Switch Assembly Life Test Example

Weibull slope represents the shape of the cumulative failure distribution. A Weibull slope less than one signifies infant mortality, a slope equal to one represents random failures, and a slope greater than one indicates wear-out characteristics (see Figure 6.9).

7. Draw a horizontal line from the 62.3 percent level on the ordinate axis to the "best fit" line from the manual approximation approach. Continue from this point and draw a vertical line down to the abscissa axis. Read the time or cycle to failure numerical value at the intersection of the vertical line to the abscissa (the software program "Weibull Smith" automatically determines the characteristic life). This corresponding value is equivalent to the characteristic life θ of the failure distribution and represents the operating duration (cycles, time, etc.) for 63.2 percent of the cumulative failures. The higher the characteristic life for a given Weibull slope (Weibull slope shifted to the right) the greater the reliability.

8. Determine the reliability either by reading from the graph or calculating it from a two-parameter reliability model equation. The graphical interpretation of reliability involves identifying the mission period on the x-axis, drawing a vertical line to the "best fit" line, and then extending a horizontal line to the cumulative percent failures axis. The corresponding value can then be subtracted from 100 percent in order to yield a reliability.

The calculation approach for the two-parameter reliability can be found by applying the Weibull distribution parameters to the following equation:[11]

$$R(t) = e^{-\left(\frac{t}{\theta}\right)^{\beta}}$$

(43)

where: $R(t)$ = reliability as a function of mission period t
        t = mission or operating period (time, cycles, etc.)
        θ = characteristic life (time, cycles, etc.)
        β = Weibull slope

For the three-parameter Weibull case the minimum life estimation is not zero. Graphically determining the three-parameter Weibull distribution would involve the same steps as for the two-parameter approach except that the point in which the "best fit" line crosses the zero point on the abscissa axis indicates a minimum life δ. The reliability can be approximated from the graph as described before or calculated from the three Weibull distribution parameters which were determined by the graphical approach:[11]

$$R(t) = e^{\left(\frac{t-\delta}{\theta}\right)^{\beta}}$$

(44)

where: $R(t)$ = reliability as a function of mission period t
        t = mission or operating period (time, cycles, etc.)
        θ = characteristic life (time, cycles, etc.)
        β = Weibull slope
        δ = minimum life (time, cycles, etc.)

9. Determination of confidence limits (one or two-sided) to the best fit representations for the Weibull plots is possible by identifying the percentile value to each ordered data value from a median ranks table. Once these percentile values are determined from the median ranks table they are plotted on the Weibull graph and connected with a smooth curve approximation.

Benefits of the Method

The Weibull distribution approach to reliability analysis provides many benefits to the analyst:

1. Allows for the determination of key parameters describing the failure mode distribution.
2. Provides a methodical approach to determining the reliability based on the parameters from a failure mode distribution.
3. Enables the analyst to compare results from life tests in a graphical manner.
4. Provides statistical confidence limits to the Weibull plot.
5. Enables quick and easy interpretation of life test results in a graphical manner. The Weibull distribution approach allows for graphical comparison of several data distributions.
6. Helps to provide parameters of a failure mode distribution for use in a mathematical calculation of reliability.
7. Provides a measure of reliability verification or demonstration and a graphical view of the degree of demonstration relative to a reliability requirement.

B. Reliability Test Data Evaluation

4. System Reliability Block Diagram Analysis

What and Why the Tool Is Applied

System reliability block diagram analysis involves a systematic determination of total system reliability through the calculation of component series/parallel reliability relationships from a functional block diagram. The key to this tool is to combine reliability analyses of components and subsystems or individual process operation reliabilities to better understand and forecast the functional performance behavior of system designs and processes early during product/process development. This form of analysis is important to engineering from the standpoint of gaining insight on system risk as a function of environmental stresses. Just like design reliability, total process reliability can be determined by evaluation of the series reliability or probability of success for the individual process operations. In addition, the analysis allows careful examination of the system configuration and possible alternatives before serious financial investment is made on building prototype samples and testing.

When the Tool Is Applied

System reliability block diagram analysis should be performed during the early stages of the product/process development phase of the PDP. The execution of such an analysis is contingent on the obtainment of information regarding the reliabilities of the components and subsystems which comprise the system under evaluation. If information on previous components and subsystems is available, then the system reliability study may be performed to gain some initial insight on predicted reliability and allow consideration to other design alternatives. Total process reliability can be investigated early during process checkout in order to understand the probability of success at each process operation.

Where the Tool Is Applied

The reliability block diagram analysis approach is most beneficial when applied to evaluate new or modified designs. It is particularly useful when deciding which design alternative has the best reliability advantage. The determination of total process reliability can be completed on new, modified, or carryover assembly line operations.

Who Is Responsible for the Method

The responsibility of carrying out the analysis would involve the reliability or quality engineer with support from the product design and process engineer. The design and process engineers would need to define the product design and process in as much detail as possible and provide any knowledge of previous performance.

How the Tool Is Implemented

The procedure for conducting a system design reliability analysis using block diagrams involves the construction of a reliability block representation based on the functional block diagram for the system, and a calculation of total reliability based on the mathematical representation of the individual part or subsystem reliabilities. This technique is highlighted as follows:

1. Construct a functional block diagram of the proposed design (see Figure 5.1). This functional block diagram is basically a representation of the complete design including functional elements, interfaces, and the directional relationships between the elements. The development of the functional block diagram should involve participation between various engineering groups including product design, process, and product assurance engineering personnel. It is essential that the product assurance engineer understands the nature of failure and its relationship to the component and subsequent assembly operation.

2. Transform the subsystem or system into a series, parallel, or complex series/parallel diagram based on the nature of failure of each component or subsystem on the functionality of the system. This diagram should show the effect on the system when a component or subsystem fails. The total system reliability will be expressed as a function of the series/parallel configuration (see Figure 5.2). The relationship between a series and parallel reliability configuration and their mathematical representations are illustrated in Figure 6.12.

3. Identify or determine the failure rate model based on a failure rate distribution (exponential, normal, etc.) for each component or subsystem of the reliability block diagram.

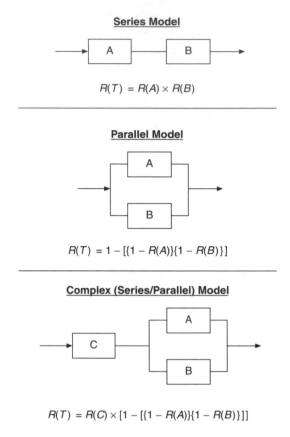

Series Model

$$R(T) = R(A) \times R(B)$$

Parallel Model

$$R(T) = 1 - [\{1 - R(A)\}\{1 - R(B)\}]$$

Complex (Series/Parallel) Model

$$R(T) = R(C) \times [1 - [\{1 - R(A)\}\{1 - R(B)\}]]$$

Figure 6.12   Basic Types of Reliability Block Diagrams and Their Mathematical Representations

This information may be gathered from warranty and field data, or derived from empirical models based on test data or engineering simulation. These models should include environmental stress factors (temperature, vibration, etc.) to which the system is exposed during its useful life (see example in Figure 6.13).

4. Gather information on the mission time equivalent to the proposed target period. This mission period may be expressed as a specific percentile level of customer usage over an established period (for example, the 95th percentile customer usage of vehicle door slam cycles for the 100,000-vehicle-mile mission period).

5. Calculate the reliability for each of the components and subsystems in the system level block diagram. These calculated reliabilities should include parameters of the defined distribution and the mission time (see example in Figure 6.13).

6. Calculate the total reliability by computing the resultant from the series/parallel mathematical model for the complete system (see Figure 6.14). Other output reports can be generated from the system reliability block diagram analysis such as failure rate or reliability versus temperature, etc. (see Figures 6.14 and 6.15).

The procedure for conducting a total process reliability analysis using the block diagram approach basically involves forming the process flowchart, generating the mathematical model for the total process probability of success, determining the probability of success for each operation, and calculating the total process probability of success from the model. Details of this procedure are highlighted in the following list:

1. Develop or modify a process flow diagram to illustrate the sequence of process operations necessary to build the product. Include only those operations involved in the build of the

A. *DC Motor Armature Reliability Calculation*

The DC motor armature portion of the total assembly has a failure rate determined by the following model:[12]

λ_{AR} = F (bearing 4, armature shaft, windings, laminations, and commutator segments) and associated internal armature connection interfaces

$$\lambda_{AR} = \left(\frac{t^2}{\alpha_1^3} + \frac{1}{\alpha_2} \right)$$

where: $t = 2000$ hours: average motor operating life (hrs.)

The Weibull characteristic life (hrs) for the motor ball bearing, shaft, and associated internal armature connection is expressed as α_1:

$$\alpha_1 = \left(10^{\left(2.534 - \frac{2357}{T+273} \right)} + \frac{1}{10^{\left(20 - \frac{4500}{T+273} + 300 \right)}} \right)^{-1}$$

Given $T = 85°C$ for the ambient operating temperature:

$$\alpha_1 = 11{,}200$$

The Weibull characteristic life (hrs) for the winding, laminations, and commutator segments combined is expressed as α_2:

$$\alpha_2 = 10^{\left(\frac{2057}{T + 273} - 1.83 \right)}$$

Given $T = 85°C$ for the ambient operating temperature:

$$\alpha_2 = 5.70 \times 10^4$$

The total assembly failure rate can be calculated using α_1, α_2 and t:

$$\lambda_{AR} = 20.3910 \text{ failures}/10^6 \text{ hrs.}$$

The DC motor armature reliability can be determined based on the following equation:

$$R_{AR}(t) = e^{-\lambda_{AR}t}$$

Total Reliability Designation	50th Percentile	95th Percentile
R_{AR}(mission #1 period)	432 hrs. = 0.991	552 hrs. = 0.989
R_{AR}(mission #2 period)	1800 hrs. = 0.964	2300 hrs. = 0.954
R_{AR}(mission #3 period)	3600 hrs. = 0.929	4600 hrs. = 0.910

Notes:
1. The failure rate model is a function of ball bearing 4, armature shaft, windings, laminations, and commutator segments. Particular to DC motor applications, the bearings and windings contribute to a majority of the failure modes and become mathematically weighted more than the other components in the armature for the failure rate model.
2. The instantaneous failure rates experienced by motors are not constant, but increase with time. The failure rate model used in this prediction is the average failure rate for the motor operating for the time period, $t = 2000$ hrs.

Continued.

Figure 6.13 System Reliability Block Diagram Analysis—DC Motor Example

3. The failure rate model is based on ball bearings with a grease lubricant. Self-lubricating bearings are used in this application. A direct correlation does not exist; however, the model is useful for the purpose of estimation.

B. *System Reliability Block Diagram Model*

The reliability of the DC Motor Assembly can be defined by the following expression (see the reliability block diagram in Figure 5.2):

$$R_{DC} = R_{BH1} \times R_{BH2} \times R_{AR} \times R_{BS1} \times R_{BS2} \times R_{MG} \times R_{AC} \times (1 - (1 - R_{CN1})$$
$$(1 - R_{CN2})) \times (1 - (1 - R_{CN1})(1 - R_{CN2})) \times R_{BR1} \times R_{BR2} \times R_{BR3} \times$$
$$R_{OG} \times R_{PG} \times R_{BP} \times R_{MH} \times R_{WW1} \times R_{WW2} \times R_{GB} \times R_{GT1} \times R_{GT2} \times$$
$$R_{EP} \times R_{OS} \times R_{MR} \times R_{BD1} \times R_{BD2}$$

where: $R_{DC}(t)$ = DC motor assembly reliability at time t.
$R_{BH1}(t)$ = brush 1 reliability at time t.
$R_{BH2}(t)$ = brush 2 reliability at time t.
$R_{AR}(t)$ = armature components reliability at time t.
$R_{BS1}(t)$ = brush compression spring 1 reliability at time t.
$R_{BS2}(t)$ = brush compression spring 2 reliability at time t.
$R_{MG}(t)$ = magnet reliability at time t.
$R_{AC}(t)$ = actuator connector reliability at time t.
$R_{CN1}(t)$ = connector to brush holder crimp connector reliability at time t.
$R_{CN2}(t)$ = connector to brush holder solder connection reliability at time t.
$R_{BR1}(t)$ = ball bearing 1 reliability at time t.
$R_{BR2}(t)$ = ball bearing 2 reliability at time t.
$R_{BR3}(t)$ = ball bearing 3 reliability at time t.
$R_{OG}(t)$ = output shaft gear reliability at time t.
$R_{PG}(t)$ = pinion gear reliability at time t.
$R_{BP}(t)$ = brush mounting plate reliability at time t.
$R_{MH}(t)$ = motor housing reliability at time t.
$R_{WW1}(t)$ = wave washer 1 reliability at time t.
$R_{WW2}(t)$ = wave washer 2 reliability at time t.
$R_{GB}(t)$ = gear box housing reliability at time t.
$R_{GT1}(t)$ = O-ring gasket 1 reliability at time t.
$R_{GT2}(t)$ = O-ring gasket 2 reliability at time t.
$R_{EP}(t)$ = end plate reliability at time t.
$R_{OS}(t)$ = output shaft reliability at time t.
$R_{MR}(t)$ = magnet retainer reliability at time t.
$R_{BD1}(t)$ = brush holder 1 reliability at time t.
$R_{BD2}(t)$ = brush holder 2 reliability at time t.

C. *Reliability Calculation*

Total system reliability for the DC motor assembly can be determined based on the following equation:

$$R_{DC}(t) = e^{-\lambda_{DC}t}$$

Total Reliability Designation	50th Percentile	95th Percentile
R_{DC}(mission #1 period)	432 hrs. = 0.987	552 hrs. = 0.983
R_{DC}(mission #2 period)	1800 hrs. = 0.947	2300 hrs. = 0.933
R_{DC}(mission #3 period)	3600 hrs. = 0.897	4600 hrs. = 0.870

Figure 6.13   *Continued.*

Temp (°C)	Failure Rate Per 1E + 6 (system hrs)	95th Percentile Reliability		
		$R(t = 552$ hrs.$)$	$R(t = 2300$ hrs.$)$	$R(t = 4600$ hrs.$)$
0	97.5481	0.947578	0.799027	0.638444
10	11.7795	0.993519	0.973271	0.947256
20	10.1612	0.994407	0.976900	0.954334
30	10.6100	0.994160	0.975892	0.952366
40	11.5079	0.993668	0.973879	0.948440
50	13.0483	0.992823	0.970435	0.941744
60	15.2826	0.991600	0.965461	0.932114
70	19.1789	0.989469	0.956847	0.915557
80	25.5476	0.985997	0.942934	0.889124
85	30.2064	0.983464	0.932884	0.870272
90	36.9081	0.979833	0.918615	0.843853
100	59.8404	0.967508	0.871418	0.759370
110	112.8519	0.939606	0.771391	0.595045
120	239.1803	0.876317	0.576884	0.332795
130	567.8930	0.730901	0.270860	0.073365
140	1353.0069	0.473852	0.044516	0.001902

Figure 6.14 System Reliability Analysis Summary Chart—DC Motor Example

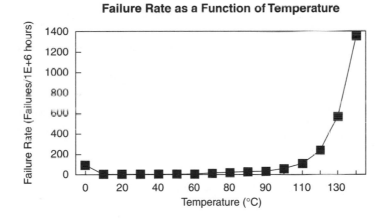

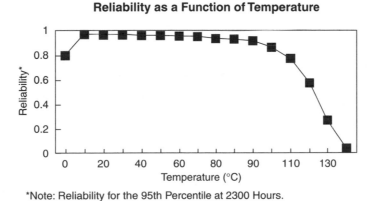

*Note: Reliability for the 95th Percentile at 2300 Hours.

Figure 6.15 Temperature Impact Profiles—DC Motor Example

product. Do not include those operations which do not directly affect the build (inventory, transfer of incoming material, etc.).

2. Construct a mathematical model which reflects the series, parallel, or series/parallel relationship of a process failure for the manufacturing build sequence. Most processes will emulate a series reliability model.

3. Determine the process defect distribution and probability of success (process reliability) at each process operation based on the attribute or variable data that is collected. If variable data is collected on process variation, then examine the process performance or capability of the process parameter and calculate the probability of success for that particular data distribution. If attribute data is collected on inspected defective parts, then determine the proportion defective from the total produced at that process operation.

4. Incorporate all the individual process operation probabilities of success into a mathematical model for the block diagram (similar to design reliability previously discussed) and compute the total process probability of success.

Benefits of the Tool

The key attributes for the application of system reliability block diagram analysis involve the ability to examine trade-offs of the total system given the various design or process configurations and known component or subsystem reliability performance. Highlights of other benefits for system reliability block diagram analysis include the following:

1. Provides the opportunity to evaluate other design or process system alternatives and examine the trade-offs in reliability.

2. Promotes an understanding of where in a system design or process the main contributors to unreliability exist.

3. Generates attention to the determination of individual system design component or process failure rates. It prompts interest on how the design or process can be improved.

4. Forces attention to the method of determining the failure rates of components and subsystems or probability of success for various assembly operations.

5. Helps establish a reliability model as a basis for design and process improvement.

C. Quality Engineering Evaluation

1. Process Performance and Capability Analysis

What and Why the Tool Is Applied

Process performance studies involve the use of statistical measures to provide an estimation of process variation and location of test sample measurements compared to specification limits. This kind of study may be performed on variable and attribute type data. These statistical indicators are used to help the analyst understand the degree of variation in product design and process characteristics and provide the basis for product design and process improvement. In addition, these indices are especially important in determining whether a process is stable, capable, and on-target.

The performance indices used for variable type data are P_p, P_{pk}, and a measure of process on-target, $P_p - P_{pk}$. The performance index P_p is used to indicate the spread of the measurements as compared to the spread of tolerance or specification limits. The performance index P_{pk} is an indicator of the measurement sample distribution location and spread relative to the closest specification limit. The process on-target index $P_p - P_{pk}$ is an indicator of how well the measurement distribution is centered to the nominal specification. Regarding attribute type data, the process potential is the only index examined and is determined by calculating the probability of success of each part evaluated per some acceptance standard. Each of these indices evaluated collectively for a sample distribution facilitates the interpretation of product design and process characteristic behavior.

Process capability studies involve similar statistical indices as process performance studies. The key difference between the two types of indices is that process capability relates to statistically stable processes where as process performance does not. Process performance studies involve a preliminary view of process capability and a calculation of the statistical indices for a smaller sample size of parts. Process capability indices C_p, C_{pk}, and C_p–C_{pk} are interpreted the same as for the process performance indices. The only difference in the calcultion of these indices appears in the estimation of the standard deviation.

When the Tool Is Applied

Process performance studies are first conducted during the product design and process development phase and evaluated, to a large extent, in the validation phase when more samples are available. These statistical measures can be applied successively throughout the product development cycle to better understand the relationship of the sample distributions to the specification. Process performance indices are often used to assess manufacturing process parameters, but are just as useful in evaluating design parameters of sample distributions during development or varification testing.

Process capability studies are usually conducted during the production and continuous improvement phase of the product development process. As more parts are manufactured during the production phase a better idea of process variation can be assessed. These process indices are applied for statistically stable processes. The determination of process stability occurs when process control has been established through the use of statistical process control charts. These control charts indicate when a process characteristic is outside of the control limits or following a natural or unnatural process variation behavior. The assessment of the cause to unnatural variation, or out-of-control and its resolution, is essential in the control chart technique.

Where the Tool Is Applied

Process performance and capability studies are applied to any design or process complexity level (component, subsystem, or system level) and part status (new, modified, or carryover).

Who Is Responsible for the Method

The principal parties involved in the application of the tool would primarily be the product quality/reliability engineers and the manufacturing engineer. Tooling and equipment engineering would also add support in order to help define the setup and ensure the maintenance of the equipment. The quality engineer would help the process engineer evaluate the initial process run data during process development as well as help interpret and initiate product improvement during production. The quality and reliability engineers may use these analysis techniques during product design development and validation.

How the Tool Is Implemented

Process performance and capability studies for analyzing variable or attribute type data are performed by: (1) establishing the measurement systems on those key product design and process parameters, (2) collecting the data, (3) calculating the descriptive statistics for the performance and capability indices, and (4) interpreting the results per process operation or test environment condition. Once the process performance or capability indices are determined, a root cause analysis is performed on the design or processing equipment/operation to investigate the cause behind the unfavorable process performance or capability. Determining the root cause may involve evaluating the results obtained from other studies such as measurement capability, preventive maintenance, simulation, etc. The following is a set of steps describing the technique (see Figure 6.16):

1. Establish or agree on a sufficient sample size of product to be processed per the designated production process assembly sequence. Determine the critical process parameters and identify whether the information to be collected is variable or attribute type data.

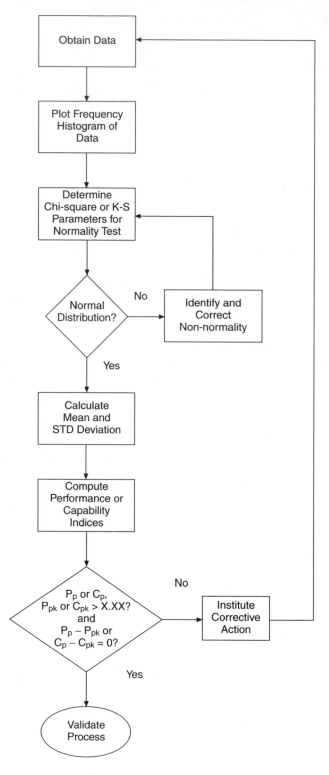

Figure 6.16 An Approach for the Analysis of Variable Data Process Performance or Capability

2. Ensure that the equipment or tool for each process operation has been calibrated and has undergone a gage repeatability and reproducibility study. Identify the calibration and total GRR of the measurement system. Measurement system variation can influence the degree of variation in the measured product design or process parameter.
3. Gather any other information about the preventive maintenance history or tool and equipment life. This information may also contribute to overall process operation variation or failure.
4. Assemble the product and collect data on measured critical design/process parameters (variable type data) and the proportion of samples built successfully per operation (attribute type data). The selection of the critical parameters for product acceptance should be compliant to the process control plan.
5. Determine the descriptive statistics of each measured product/process parameter distribution. From these statistics a determination of whether the sample distribution represents a normal distribution should occur. The use of the chi-square test for normality may be used on sample sizes of 25 or more, while use of the Kolmogorov-Smirnoff (KS) test may be used on sample sizes of 25 or less. Tests for normality should be performed to validate the type of distribution prior to evaluating the process performance or capability. The use of the process performance and capability indices is contingent on a normal distribution.

    The application of the chi-square method for determining normality is reviewed as follows (from equation #31):[13]

$$\chi^2 = \sum_{i=1}^{K} \left[\frac{\left[f(o) - f(e) \right]^2}{f(e)} \right]$$

where: χ^2 = chi-square test statistic
$f(o)$ = observed class frequency in sample
$f(e)$ = expected class frequency, assuming H_o is true

If $f(o)$ and $f(e)$ differ significantly, the test statistic will be large. H_o will be rejected (distribution is not normal) if the test statistic exceeds the critical value found in a χ^2 distribution table (see example study in Figures 6.17a and 6.17b).
6. Calculate the process performance or capability index for the predetermined product/process parameter normal distribution. The higher the value of P_p and C_p the smaller the sample distribution spread compared to the tolerance (see Figure 6.18). Similarly, the higher the value of P_{pk} and C_{pk} the closer the sample distribution is to the mean relative to the tolerance limits (see Figure 6.19). As for the measure of being on-target, the smaller the difference between P_p and P_{pk} or C_p and C_{pk} the more centered the sample mean is to the specification nominal (see Figure 6.20).

Process Performance Indices: Variable Data[14]

- Process performance indice P_p is calculated as follows:

$$P_p = \frac{(USL - LSL)}{6\sigma_s}$$

(45)

where: P_p = process performance indice
USL = upper specification limit
LSL = lower specification limit

Characteristic: Housing thickness

Sample Size = 50

Mean = 1.6189	Min. Value = 1.615	Interval = 0.0010
Sigma indiv. = 0.0015	Max. Value = 1.622	Lower boundary
Est. Sigma = 0.0013	Kurtosis = 0.169	Chi-Square = 4.913
Coeff. Var. = 0.0009	Skewness = −0.649	Deg. Free = 5
		Conf. Level = 95%
		Normal*

Actual %	*Capability (using Sigma)*	*Theoretical %*
Above Spec = 0.00	Upper Spec = 1.6250	Above Spec = 0.00
Below Spec = 0.00	Nominal = 1.6150	Below Spec = 0.00
Out of Spec = 0.00	Lower Spec = 1.6050	Out of Spec = 0.00
C_{pk} = 1.33	Cr = 0.46	Z upper = 3.99
C_p = 2.19		Z lower = 9.14
Mean + 3s = 1.6235		
Mean − 3s = 1.6143		

*Distribution is representative of a normal distribution since the chi-square test statistic value did not exceed the critical value at the 95% confidence level for 5 degrees of freedom (4.913 < 11.071). The degrees of freedom ($d.f. = k - m - 1$) was based on a number of frequency distribution classes, $k = 8$, and number of statistical parameters estimated (mean and standard deviation), $m = 2$. As a result of this normality test, the capability statistics C_p and C_{pk} can be interpreted for the normal distribution.

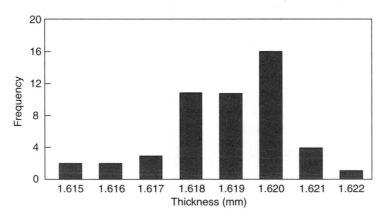

Frequency Histogram of Housing Thickness Dimension
A Determination of Distribution Normality

Figure 6.17a   Process Capability Study Based on Determination of Normality—Example

Characteristic: Bobbin inside width

Sample Size = 50

Mean = 0.51282	Min. Value = 0.5100	Interval = 0.00080
Sigma indiv. = 0.00170	Max. Value = 0.5155	Lower boundary
Est. Sigma = 0.00093	Kurtosis = −1.245	Chi-Square = 16.703
Coeff. Var. = 0.00331	Skewness = 0.222	Deg. Free = 4
		Conf. Level = 95%
		Not Normal*

Actual %	*Capability (using Sigma)*	*Theoretical %*
Above Spec = 0.00	Upper Spec = 0.52000	Above Spec = 0.00
Below Spec = 0.00	Nominal = 0.51000	Below Spec = 0.00
Out of Spec = 0.00	Lower Spec = 0.50500	Out of Spec = 0.00
C_{pk} = 1.41	Cr = 0.68	Z upper = 4.23
C_p = 1.47		Z lower = 4.61
Mean + 3s = 0.51791		
Mean − 3s = 0.50773		

*Distribution is not normal since chi-square test statistic value exceeds the critical value at the 95% confidence level for 4 degrees of freedom (16.703 > 9.488). The degrees of freedom ($d.f. = k - m - 1$) was based on a number of frequency distribution classess, $k = 7$, and number of statistical parameters estimated (mean and standard deviation), $m = 2$. As a result of this normality test, the capability statistics C_p and C_{pk} cannot be interpreted as a result of the non-normal distribution.

Frequency Histogram of Bobbin Inside Width
A Determination of Distribution Normality

Figure 6.17b Process Capability Study Based on Determination of Normality— Example *(Continued)*.

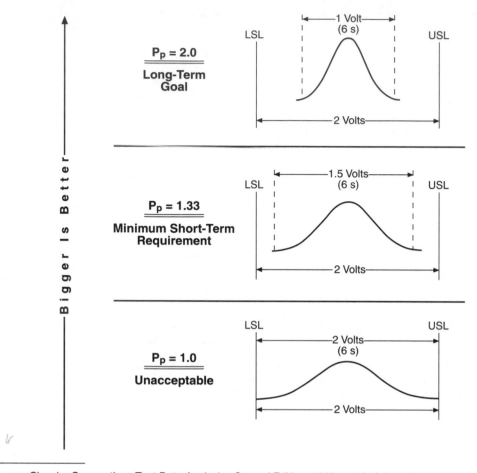

Source: Chrysler Corporation, *Test Data Analysis*—Second Edition, 1995, p. 6A. Adapted by permission.

Figure 6.18   Process Performance, P_p (Indicator of Spread)

$$s \text{ or } \sigma_s = \text{sample standard deviation} = \sqrt{\frac{\sum_{i=1}^{n}\left(x_i - \bar{x}\right)^2}{n-1}} \qquad (46)$$

$$\bar{x} = \text{sample average} = \frac{\sum_{i=1}^{n} x_i}{n} \qquad (47)$$

x_i = individual readings
n = total number of all the individual readings

- The process performance indice P_{pk} is calculated as follows:

$$P_{pk} = \text{the smaller of} \dots \frac{\left(USL - \bar{x}\right)}{3\sigma_s} \text{ or } \frac{\left(\bar{x} - LSL\right)}{3\sigma_s} \qquad (48)$$

- The "on-target" indice is calculated as follows:

$$\text{on-target indice} = P_p - P_{pk} \qquad (49)$$

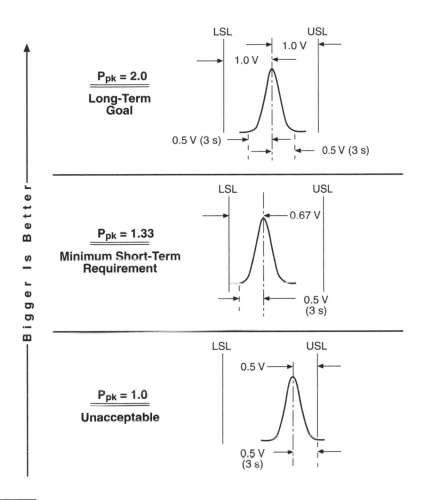

Source: Chrysler Corporation, *Test Data Analysis*—Second Edition, 1995, p. 8A. Adapted by permission.

Figure 6.19   Process Performance, P_{pk} (Indicator of Location)

Process Capability Indices: Variable Data[14]

- The process potential indice C_p is calculated as follows:

$$C_p = \frac{(USL - LSL)}{6\sigma_{\bar{R}}/d_2} \tag{50}$$

where: C_p = process potential indice
   USL = upper specification limit
   LSL = lower specification limit

$$\frac{\sigma_{\bar{R}}}{d_2} = \frac{\bar{R}}{d_2} = \text{sample process standard deviation}$$

$\bar{R}$ = average of the moving ranges
d_2 = a constant varying by the subgroup sample size (Table 7a)

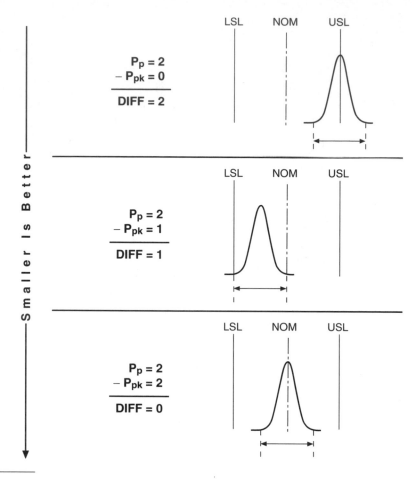

Source: Chrylser Corporation, *Test Data Analysis*—Second Edition, 1995, p. 10A. Adapted by permission.

Figure 6.20   "On Target", $P_p - P_{pk}$ (Indicator of Nominal Location)

- The process capability indice C_{pk} is calculated as follows:

$$C_{pk} = \text{the smaller of} \ldots \frac{(USL - \bar{x})}{3\sigma_{\bar{R}}/d_2} \text{ or } \frac{(\bar{x} - LSL)}{3\sigma_{\bar{R}}/d_2} \qquad (51)$$

- The "on-target" indice is calculated as follows:

$$\text{on-target indice} = C_p - C_{pk} \qquad (52)$$

Process Performance Index: Attribute Data

The determination of the probability of success (or process potential) for each operation based on the acceptance criteria of good versus bad parts:

- The probability of success index for evaluating attribute type data is as follows:

$$P_a = \text{probability of success (\%)} = \frac{\text{quantity of acceptable parts}}{\text{total quantity of parts attempted}} \times 100\% \qquad (53)$$

7. Evaluate the process performance or capability results for variable and attribute type data of all process operations in the manufacturing process. Identify which process operations

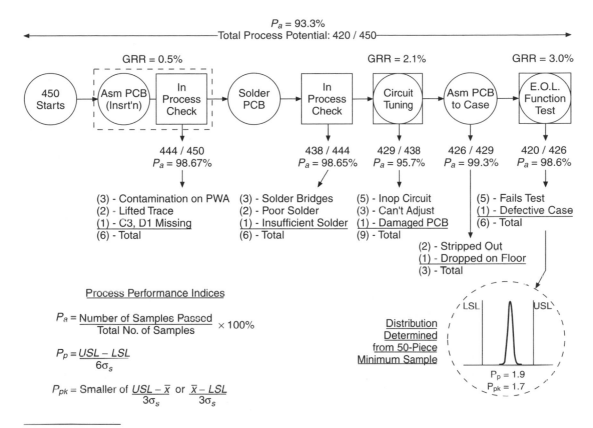

Figure 6.21 Process Performance Evaluation Example

produced low process performance or capability and low probability of success (see example study in Figure 6.21).

8. Compare those process operations with low process performance or capability and low probability of success to other analyses of the process which may suggest sources of variation (gage calibration, gage repeatability and reproducibility, tool and equipment maintainability studies, etc.). Determine if there is a relationship between results obtained from these sources and the low process performance or capability.

9. Institute adjustments and improvements to the particular process operation in order to improve the process performance or capability and probability of success. Repeat the process study by assembling more parts through the improved process and evaluating performance or capability and probability of success.

Benefits of the Tool

Process performance and capability studies help provide a measurable value for the magnitude of product design/process control. These mathematical process performance and capability indices provide a statistical understanding of the relationship of the measured or visually inspected product/process characteristic to a specification or tolerance. Some of the key benefits are:

1. Provides a statistically based method for benchmarking the variability or probability of success of a product design/process key characteristic.

2. Enables the analyst and the engineering team to evaluate and forecast the variation for each parameter of a process operation or a design under a given test environment.
3. Helps to gauge the total process potential of a product prior to conditions present during production.
4. Facilitates the communication between product design and manufacturing personnel through actual analysis.
5. Helps support any continuous improvement in a product design or process through actual data analysis before improvements or redesigning efforts are carried out.
6. Builds confidence in the ability to enter the production phase of the PDP with less risk of experiencing internal or external failures.
7. Provides a step-by-step approach to assessing process performance and capability, regardless of the data type.

C. Quality Engineering Evaluation

2. Design of Experiments (DOE)

What and Why the Tool Is Applied

The design of experiment (DOE) technique is used to understand the conditions of a product or process design such that it functions with limited variability despite the diverse and changing effects of the environment, wear, process variations, or component-to-component variations. Design of experiments is a technique used to develop and analyze the results of an experiment for determining the relationship of particular design or process factors or input variables to an output response variable. The approach facilitates the analyst's understanding of those factors or input variables in a design or process which influence the behavior of the output response variable and allows for better decisions on the type of adjustment or improvement necessary. This tool provides more detailed root cause effort for understanding product/process variation or failure and for seeking a more robust product design or process.

When the Tool Is Applied

The design of experiments is most appropriately applied during product design and process development. Using DOE as a developmental tool provides the ability to explore the sources of variation and implement product design or process adjustments in order to control the performance of the product under the given environmental condition (test environment or manufacturing process operation condition). This analytical tool is also used during the validation and production phase of the PDP. The traditional application of DOE has been during production, but can be more effective if applied during development when there is more time to study the design or process under less stressful conditions.

Where the Tool Is Applied

Design of experiments is applied to any design or process complexity level (component, subsystem, or system level) and part status (new, modified, or carryover).

Who Is Responsible for the Method

The individuals responsible for applying this technique are the quality, reliability, or product assurance specialist and the process or manufacturing engineer. The quality, reliability, or product assurance engineers would use DOE throughout the PDP, depending on the product design or process problem under investigation. This tool may even be used by product or development engineers who are investigating design performance under accelerated test conditions.

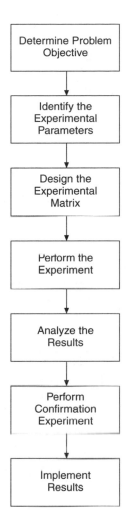

Figure 6.22   Design of Experiment Process Flowchart

How the Tool Is Implemented

The DOE process is a very useful tool in helping to investigate the factors and the level of the particular factors which affected the observed response. It basically involves establishing the test matrix containing samples with various test factor combinations and the analysis of such a matrix to identify the most significant factor to the response variable. Highlights to the steps involved in completing the development and analysis of an experiment are as follows (see Figure 6.22):

1. Identify and gather information about the problem, including conditions surrounding the problem (environmental factors, processing behaviors, etc.). Understand the problem clearly and identify a problem statement.
2. Organize the experiment by selecting a cross-functional team (product, development, process, quality, etc.) to help choose the measurable response parameter and those factors which might have contributed to an observed response leading to the problem. The selection of the design factor (dimensional, material, orientation, electrical, etc.) or process factor (temperature, speed, pressure, etc.) and the level of each factor is a critical step in the experimental process. This selection may be a result of previous studies (cause and effect, Ishakawa diagram, previous experiments, etc.) and observations (Pareto diagrams).

3. Design the experimental test matrix by choosing the classical or Taguchi orthogonal array approach and the analysis technique to evaluate the results from such a matrix. The classical approach simply involves determining the *total number* of test combinations possible for evaluating the influence of the factor type and factor levels on the output response. The Taguchi approach involves selecting the orthogonal array design, which covers a *smaller sample of the total number* of test combinations than the classical design, and evaluating the influence of the factor type and factor levels on the output response.

   The analysis of the DOE matrix may involve the construction of a means and log of the variance table (Figure 6.23) or an analysis of variance (ANOVA) table (Figure 6.24). The ANOVA table is a statistical tool for evaluating the significance level of the F-test statistic for each factor or an interaction of factors.

   The results from a means and variance or ANOVA table can be displayed using a line graph that shows the relationship of the means or log of the variance versus the factor type and level. These plots help to visualize the magnitude of influence that the various factor/levels have on the chosen output response.

4. Run the experiment giving careful thought to hardware procurement and experiment execution. Ensure that the test treatment combinations are run in a randomized order for the purpose of minimizing any possible biases. In addition, be aware of any contingency plan if the experiment is interrupted or other distractions take place.

5. Analyze the data gathered from the test through the use of means and log of variance or ANOVA tables. Results obtained should be compared between the various replications and use of different experimental error or risk levels. From the analysis it is possible to select the significant factors and optimum levels which cause significant variation on the chosen output response. This information becomes useful input to understanding those conditions which contribute to performance variation and possible failure under applicable test environments.

6. Confirm the results from the designed experiment by repeating the experiment. Possible causes for nonrepeatable results may be that a key factor was not included, an interaction was ignored, or measurement error was present.

7. Implement the results of the experiment by improving the design/process or considering alternative approaches.

Benefits of the Tool

Design of experiments provides an effective means in developing test matrices for exploring the relationship and contribution of various design or process factors to a specified output. The tool is very helpful in quantitatively understanding the effect of many factor types and levels of each factor on the test sample. A summary of these benefits is as follows:

1. Provides a statistical analysis of the factors which may contribute to the variability of the output response. The DOE process helps to address the contribution and significance level of each factor or interaction of factors.

2. Enables the engineer the flexibility of applying the test matrix and analysis approach of DOE during development, validation, and the production phase of the PDP.

3. Utilizes either the classical or a Taguchi test matrix design.

4. Can be used along with accelerated testing techniques to derive product design test sequences for development and verification testing.

5. Produces a complete analysis of variance which can be graphically displayed through the illustration of the mean or variance effect as a function of the factor level combination. This graphic illustration is quite useful to the product or process engineer who needs to interpret the results quickly and apply the results to a corrective action.

Experimental Matrix and Data Table

Log s²	Avg.	Response				A B C D	tc
−0.421	16.81	16.17	17.13	16.44	17.51	1 1 1 1	1
0.713	16.94	18.80	19.01	14.81	15.15	1 1 2 2	2
0.714	11.32	13.10	13.27	10.38	8.55	1 2 1 2	3
−0.059	12.19	11.57	12.92	11.21	13.05	1 2 2 1	4
0.961	24.09	23.87	24.55	27.66	20.29	2 1 1 2	5
1.149	23.88	20.18	25.48	21.48	28.38	2 1 2 1	6
1.175	18.60	19.17	13.51	18.80	22.92	2 2 1 1	7
0.846	18.63	16.92	16.05	21.84	19.71	2 2 2 2	8

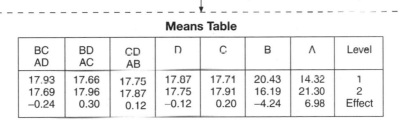

Means Table

BC AD	BD AC	CD AB	D	C	B	A	Level
17.93	17.66	17.75	17.87	17.71	20.43	14.32	1
17.69	17.96	17.87	17.75	17.91	16.19	21.30	2
−0.24	0.30	0.12	−0.12	0.20	−4.24	6.98	Effect

Note: Effect = Level 2 − Level 1

Log s² Table

BC AD	BD AC	CD AB	D	C	B	A	Level
0.332	0.572	0.580	0.461	0.608	0.600	0.237	1
0.939	0.699	0.691	0.809	0.662	0.670	1.034	2
0.607	0.127	0.111	0.348	0.054	0.070	0.797	Effect

Plot of Significant Effects

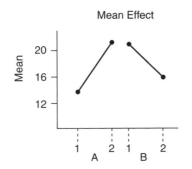

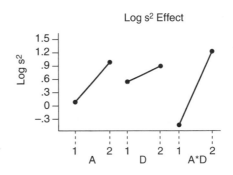

Source: John Bieda, "A Practical Product/Process Reliability Development and Improvement Approach," SAE Technical Series #932881, 1993, p. 63. Adapted by permission.

Figure 6.23 Design of Experiment Analysis Example

Design of Experiment Matrix (L8–3)

Run #	Factor A	Factor B	Factor C	Output Response
	(spring radius) 0 = 0.50 in. 1 = 1.00 in.	(spring rate) 0 = 4.0 in/lb 1 = 8.0 in/lb	(spring contact angle) 0 = 50 rad. deg 1 = 100 rad. deg	(standard deviation of spring cycles to failure)
1	0	0	1	311824
2	0	1	0	534573
3	0	1	1	6717514
4	1	1	1	5056167
5	1	0	1	957776
6	1	1	0	44548
7	1	0	0	2672864
8	0	0	0	66822

Analysis of Variance (ANOVA) Table

Factor/Combination	Sum of Squares	DF	Mean Square	F-ratio	P-Value
A	1.514E+11	1	1.514E+11	1.95	0.3898
B	8.702E+12	1	8.702E+12	111.89	0.0592
C	1.182E+13	1	1.182E+13	151.99	0.0508
AB	3.649E+12	1	3.649E+12	46.93	0.0910
AC	1.226E+12	1	1.226E+12	15.76	0.1549
BC	2.005E+13	1	2.005E+13	257.80	0.0391
Total Error	7.777E+10	1	7.777E+10		
Total (corr.)	4.568E+13	7			

R-squared = 0.998297   R-squared (adusted for DF) = 0.988081

Spring Assembly Designed Experiment
Standardized Effects of Spring Cycles Standard Deviation*

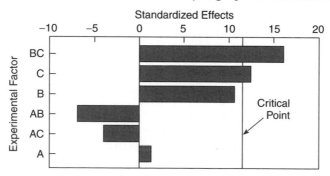

*Note: The chart includes a vertical line at the critical T-value for an alpha = 0.05 (95% confidence level).
An effect that exceeds the vertical line may be considered significant.

Figure 6.24   Design of Experiment—Mechanical Spring Assembly Example

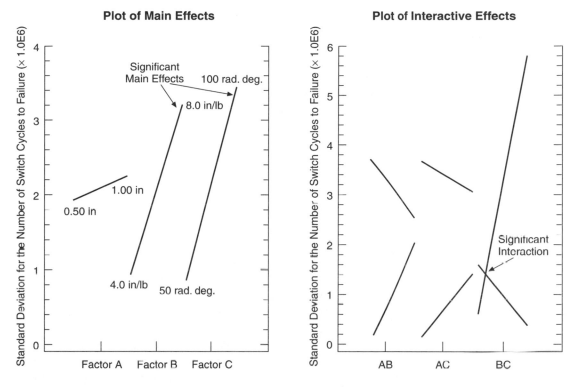

Figure 6.24 *Continued.*

6. Complements the root cause approach in helping to investigate the mechanisms behind variability or failure in a product design or process.
7. Helps save time and effort in determining a test design that will cover the possible combinations of test conditions in one test matrix.
8. Provides an organized approach to the test development and review process. The DOE process helps to generate results which can be evaluated and presented in a simple manner to the engineering team in a design review.

D. Process Review

What and Why the Tool Is Applied

The process review is a formal and preliminary evaluation of a production assembly process and its tooling based on several key elements. These elements form the basis for operation, support, and maintenance of an assembly process. In particular, the process review involves a documentation examination and a facility visit for manufacturing process verification. An examination as to the presence, control, and maintenance of the elements for the process is important prior to actual production of the product design. The results from the review help to point out deficiencies, suggest improvements, and allow observation of the production system under actual production intent conditions. Conducting a review of the complete process allows an opportunity to benchmark the process prior to production. In addition, a review of these considerations addresses some of the key elements specified in QS-9000. This examination should encompass the following areas of any process:

• Process flow diagram and description
• Process FMEA

- Process control plan (identifying all design and process control characteristics)
- Incoming and outgoing material acceptance plan
- Statistical process control
- Scheduled preventive maintenance plans
- Tooling and equipment studies
- Measurement systems plan and evaluation studies
- Previous and current process performance and capability studies (by station and total process)
- Product assurance plan
- Error and mistake proofing
- Operator and machine instruction sheets with process control characteristics
- Parts handling plan
- Parts packaging and shipping plan
- Quality system for root cause analysis and corrective action
- Process verification test

When the Tool Is Applied

Process review is conducted in the middle stages of the process development phase through and including the beginning of the process verification test phase. This review should be performed prior to an actual production facility visit through the careful examination of the process review elements. The evaluation of the process review elements should be an ongoing activity during product design and process development. Demonstration of production build under actual line conditions (rate, operator expertise, etc.), acceptable process performance or capability results, and successful performance of the product design after process verification testing are important elements included in this review. The identification and demonstration of statistical process control and preventive maintenance is necessary and must be demonstrated periodically to the quality department of the reviewing party members.

Where the Tool Is Applied

The process review is applied regardless to any design complexity (component, subsystem, or system level). These reviews are mostly directed to new or modified product designs and processes. However, a periodic review of the carryover process is important since slight changes in the process operation or process variation could affect the quality and reliability of the part being produced.

Who Is Responsible for the Method

The individuals responsible for carrying out the process review are the quality engineer, product engineer, and process or manufacturing engineer. The product assurance specialist would also assist in the review of the documentation and support the review of the facility operation.

How the Tool Is Implemented

The process review would occur through the evaluation of the documentation prior to a visit to the assembly operation. This review would involve the evaluation of each of the 16 process review considerations and how they specifically address key elements in the QS-9000 Assessor's Guide (see Figure 6.25).[15] A brief summary of each process review consideration is highlighted in the following text:

1. Process Flow Diagram and Description:

    The process flow diagram would contain a sequence of process steps and a brief description of operation. A review of this document would involve clear and accurate representation of the process as it exists at the manufacturing site. The use of the process flow diagram and the information it conveys helps facilitate documentation of the quality system (procedures, work instructions, control plans, etc.).

Process Review Elements

Quality System Requirements (QS-9000 Elements)	Process Flow Diagram	Process FMEA	Process Control Plan	Incoming/Outgoing Acceptance Plan	Statistical Process Control	Scheduled Preventive Maintenance Plan	Tooling and Equipment Studies	Measurement System Plan and Evaluation	Process Performance and Capability Studies	Product Assurance Plan	Error and Mistake Proofing	Operator and Machine Instruction Sheets	Parts Handling Plan	Parts Packaging and Shipping Plan	Root Cause and Corrective Action	Process Verification
Management Responsibility (4.1)																
Quality System (4.2)	X	X	X							X						
Contract Review (4.3)																
Design Control (4.4)		X										X				X
Document and Data Control (4.5)																
Purchasing (4.6)																
Control of Customer Supplied Product (4.7)				X		X										
Product Identification and Traceability (4.8)																
Process Control (4.9)						X			X			X				
Inspection and Testing (4.10)				X												
Inspection, Measuring, and Test Equipment (4.11)								X								
Inspection and Test Status (4.12)																X
Control of Nonconforming Product (4.13)																
Corrective and Preventive Actions (4.14)															X	
Handling, Storage, Packaging, and Delivery (4.15)													X	X		
Control of Quality Records (4.16)																
Internal Quality Audits (4.17)																
Training (4.18)																
Servicing (4.19)																
Statistical Techniques (4.20)					X		X		X							

Figure 6.25 Relationship Between QS-9000 and Process Review Elements

165

Purpose

A process of evaluating purchased or manufactured product in a lot for the purpose of accepting or rejecting the entire lot for conformance per quality specification.

Application

Incoming (receiving inspection) and outgoing (AQL) inspection

Types of Sampling Plans

1. Attribute (MIL-STD-105F)—see example in Figure 6.27a
2. Variable (MIL-STD-414)

Measures of Acceptance Sampling Effectiveness

1. Operating Characteristic (OC) Curves—see example in Figure 6.27b
2. Average Outgoing Quality (AOQ) Curves
3. Lot Acceptance Sampling per Severity Class

Sampling Plan Parameters

1. Lot Size (N)
2. Sample Size (n)
3. Acceptance Number (c)
4. Severity Class (ex. minor, major, critical, etc.)

Figure 6.26 Acceptance Sampling Highlights

2. Process FMEA:

The process FMEA contains a risk assessment per assembly step and the determination of a control characteristic. A review of this document should consider the type of failure modes specified, ranking of the risks, and the identification of the current control.

3. Process Control Plan:

The process control plan describes in detail how the current control parameter, as first identified in the PFMEA, is measured, evaluated, and statistically controlled for the particular process step. A review of this document would consider correlation to the PFMEA, nature of the measurement of the characteristic, and the type of control used. A reaction plan would be implemented in the event of an out-of-control situation.

4. Incoming and Outgoing Material Acceptance Plan (see Figures 6.26, 6.27a, and 6.27b):

The incoming and outgoing material acceptance plan basically involves the identification of the sample size necessary for inspection relative to lot size and criticality of the-product design/process characteristic under inspection. A review of this plan for both incoming and outgoing material should consider the procedure and implementation. Present philosophy for acceptance sampling for defects less than 1000 PPM (parts per million) is either 100 percent or 0 percent. Alternative approaches involving sampling per design/process severity are possible plans as well.

MIL-STD-105D Master Table for Normal Inspection (Single Sampling)[16]

Acceptable Quality Levels (normal inspection). Each cell shows **Ac Re** (Acceptance number / Rejection number). ↓ = use first sampling plan below arrow; ↑ = use first sampling plan above arrow.

Sample size code letter	Sample size	0.010	0.015	0.025	0.040	0.065	0.10	0.15	0.25	0.40	0.65	1.0	1.5	2.5	4.0	6.5	10	15	25	40	65	100	150	250	400	650	1000
A	2															↓	0 1	1 2	2 3	3 4	5 6	7 8	10 11	14 15	21 22	30 31	44 45
B	3														↓	0 1	1 2	2 3	3 4	5 6	7 8	10 11	14 15	21 22	30 31	44 45	↑
C	5													↓	0 1	1 2	2 3	3 4	5 6	7 8	10 11	14 15	21 22	30 31	44 45	↑	
D	8												↓	0 1	1 2	2 3	3 4	5 6	7 8	10 11	14 15	21 22	30 31	44 45	↑		
E	13											↓	0 1	1 2	2 3	3 4	5 6	7 8	10 11	14 15	21 22	30 31	44 45	↑			
F	20										↓	0 1	1 2	2 3	3 4	5 6	7 8	10 11	14 15	21 22	30 31	44 45	↑				
G	32									↓	0 1	1 2	2 3	3 4	5 6	7 8	10 11	14 15	21 22	30 31	44 45	↑					
H	50								↓	0 1	1 2	2 3	3 4	5 6	7 8	10 11	14 15	21 22	30 31	44 45	↑						
J	80							↓	0 1	1 2	2 3	3 4	5 6	7 8	10 11	14 15	21 22	30 31	44 45	↑							
K	125						↓	0 1	1 2	2 3	3 4	5 6	7 8	10 11	14 15	21 22	30 31	44 45	↑								
L	200					↓	0 1	1 2	2 3	3 4	5 6	7 8	10 11	14 15	21 22	30 31	44 45	↑									
M	315				↓	0 1	1 2	2 3	3 4	5 6	7 8	10 11	14 15	21 22	30 31	44 45	↑										
N	500			↓	0 1	1 2	2 3	3 4	5 6	7 8	10 11	14 15	21 22	30 31	44 45	↑											
O	800		↓	0 1	1 2	2 3	3 4	5 6	7 8	10 11	14 15	21 22	30 31	44 45	↑												
P	1250	↓	0 1	1 2	2 3	3 4	5 6	7 8	10 11	14 15	21 22	30 31	44 45	↑													
R	2000	0 1	1 2	2 3	3 4	5 6	7 8	10 11	14 15	21 22	30 31	44 45	↑														

↓ = Use first sampling plan below arrow. If sample size equals , or exceeds, lot or batch size, do 100% inspection.

↑ = Use first sampling plan above arrow.

Ac = Acceptance number.

Re = Rejection number.

Sample Size Code Letters*

Lot or batch size	General inspection levels		
	I	II	III
2 to 8	A	A	B
9 to 15	A	B	C
16 to 25	B	C	D
26 to 50	C	D	E
51 to 90	C	E	F
91 to 150	D	F	G
151 to 280	E	G	H
281 to 500	F	H	J
501 to 1,200	G	J	K
1,201 to 3,200	H	K	L
3,201 to 10,000	J	L	M
10,001 to 35,000	K	M	N
35,001 to 150,000	L	N	P
150,001 to 500,000	M	P	Q
500,001 and over	N	Q	R

*Sample size code letters given in body of table are applicable when the indicated inspection levels are to be used. The Standard includes an added table of code letters for small-sample inspection.

Figure 6.27a Attribute Sampling Plan (MIL-STD-105D) Example

5. Statistical Process Control:

Statistical process control involves the use of attribute and variable data type control charts for monitoring the variability in a particular process characteristic. A review of this tool should consider adaptation of the charts relative to the control plan, analysis, and the interpretation of trends and out-of-control conditions.

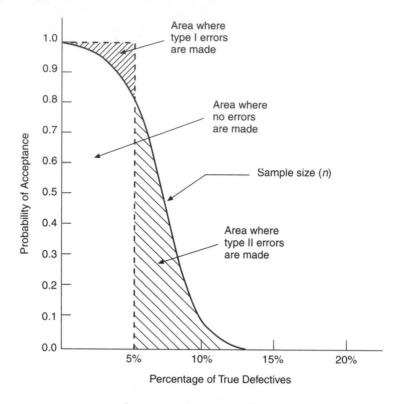

Operating Characteristics Curve

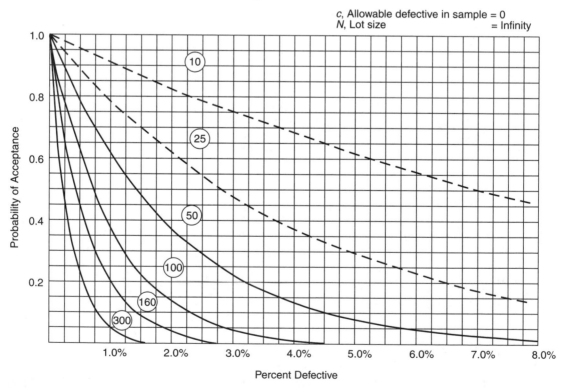

Note: (#) = Sample size

Figure 6.27b Operating Characteristic Curves

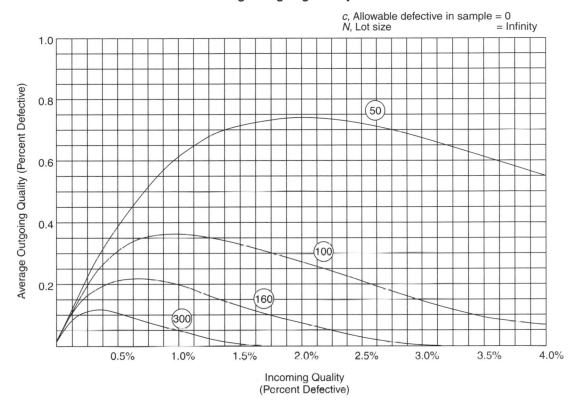

Average Outgoing Quality Curves

Note: (#) = Sample size

Figure 6.27b *Continued.*

6. Scheduled Preventive Maintenance Plan:

   The scheduled preventive maintenance plan contains a description of the kind of maintenance activity and its frequency during a production period. A review of this plan should consider the determination of the time to repair and the maintenance procedure for a particular piece of equipment.

7. Tooling and Equipment Studies:

   Tooling and equipment studies would involve the technique and analysis supporting the determination of the mean time to repair or provide a particular maintenance activity. A review of these studies should consider where and how the frequency of repair was developed as well as how it is updated.

8. Measurement Systems Plan and Evaluation:

   The measurement systems plan and evaluation activity would involve the identification and technique used for the inspection and capability of a measurement tool or piece of equipment. This review should consider calibration and gauge repeatability and reproducibility of all variable and attribute type gauges and inspection systems. Evidence and demonstration of the analyses would be necessary as well as the frequency of evaluation.

9. Process Performance and Capability Studies:

   Process performance and capability studies would involve the evaluation technique and procedure for carrying out an analysis relative to attribute and variable type data of specific product design/process characteristics. A review of these kinds of studies would

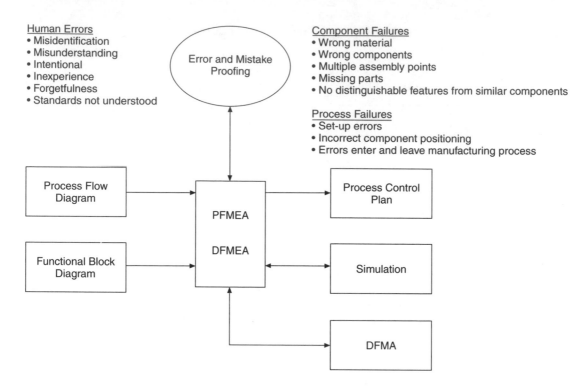

Figure 6.28 Relationship of Error and Mistake Proofing to Design/Process Development Activities

consider the identification of design/process parameter, the sampling procedure, and the analysis technique such as verifying validity of the normal distribution prior to calculation of the performance and capability indices.

10. Product Assurance Plan:

A product assurance plan would involve a schedule of events regarding engineering activities necessary to help develop and evaluate a product design and process for a specific product development process schedule. A review of this plan would consider the type of tools and methods as well as engineering activities used to accomplish the engineering milestones.

11. Error and Mistake Proofing (see Figure 6.28):

The use of error and mistake proofing techniques would involve designing a potential failure mode out of a product or process (error proofing) and any change to the operation that helps the operator reduce or eliminate mistakes (mistake proofing). A review of the use of error and mistake proofing should consider evidence of the feature in a piece of equipment or process step operation which would eliminate or prevent failure modes as identified in the PFMEA.

12. Operator and Machine Instruction Sheets:

Operator and machine instructions specifically describe the steps necessary in performing the process operation or setting up and adjusting equipment. They would include reference and directions as to the handling of key product design/process characteristics and their control. A review of the operator and machine instructions would consider the treatment of the key product design/process characteristics per the process control plan, application of visual aids to detect anomalies, and ability to follow the directions and perform the operation without difficulty.

13. Parts Handling Plan:

    The parts handling plan would involve the details behind such things as containers used, transfer of material to the line and between operations, tracking of the material, etc. A review of the parts handling plan would involve specific directions on treatment of material coming in, during, and going out of the manufacturing facility, and would involve actual viewing of the operation.

14. Parts Packaging and Shipping Plan:

    The parts packaging and shipping plan for a manufacturer should include a specific design of the package, a test plan for validating the package, details behind the shipping procedure, and a plan for handling returnable containers. A review of parts packaging and shipping would involve verification of the packaging design and test plan.

15. Root Cause and Corrective Action Procedure:

    Root cause and corrective action involves documented evidence of procedures and examples on how problems in the manufacturing process are evaluated and resolved. A review of the reaction plans and the studies conducted to resolve these problems is an important element of this activity. The review root cause and corrective action would necessitate the logical process of (1) detecting the problem, (2) evaluating data about the problem, (3) performing any experiments, (4) analyzing and interpreting the results, (5) implementing the solution, and (6) a follow-up for effectiveness of corrective action, which "closes the loop".

16. Process Verification:

    Process verification testing would involve the execution of a design verification test for parts which have been produced from production tools and processes. A review of this activity would involve the analysis of the test results, including statistical studies on the variability of the part performance under environmental stress as well as the demonstrated reliability.

Benefits of the Tool

Process verification and review during design/process development provides the assurance that the production intent part is prepared for manufacture without significant risk during the production phase of the PDP. This process review activity requires significant attention to the documentation and the procedures actually in place at the manufacturing site. Some benefits of the review process include the following:

1. Provides an ability to thoroughly evaluate the process, make improvements, and see the outcome of these adjustments before launching the product into the production phase.
2. Helps to validate the practices and procedures of the manufacturing process.
3. Allows for a preliminary evaluation of the process relative to specific questions posed for each element in QS-9000. This process verification and review addresses specific guidelines as called out in QS-9000.
4. Helps the engineering team to understand the details of the process and contributes assistance and support in addressing areas of the process where significant risk is present.
5. Enables the complete engineering team to be involved in the process review.

NOTES

1. Statistical Graphics Corporation, *Statgraphics—Statistical Graphics System,* Version 4.0, 1989.
2. Adapted from F. J. Massey, Jr., "The Kolmogorov-Smirnov Test for Goodness of Fit," *Journal of American Statistical Association,* Vol. 46, 1951, pp. 68–78. Originally adapted by Bernard Ostle, *Statistics in Research,* 2nd Edition, (Iowa State University Press, 1963), p. 560.

3. American Society for Quality Control Statistics Division, *Glossary and Tables for Statistical Quality Control,* (Milwaukee, WI: ASQ, 1983), p. 101.

4. ——— pp. 97–99.

5. Chrysler, Ford, and General Motors Supplier Quality Requirements Task Force, *Fundamental Statistical Process Control—Reference Manual,* 1991, pp. 79–82.

6. Daniel, Wayne W., and James Terrel, *Business Statistics,* 4th Edition, (Boston, MA: Houghton Mifflin Company, 1986), pp. 94–104.

7. ——— pp. 108–112.

8. McCullen, Lawrence R., *RELIBISS*—A Stress/Strength Computer Software Program, Electrical Systems Center, General Motors Corporation, 1985.

9. Nelson, Wayne, "Weibull Analysis of Reliability Data with Few or No Failures," *Journal of Quality Technology,* Vol. 17, No. 3, 1985, pp. 140–146.

10. Abernathy, R. B., *Weibull Smith*—A Weibull Analysis Computer Software Program, Fulton Findings™, 1988.

11. Kececioglu, Dimitri, *PE, Lecture Notes for AME 408—Reliability Engineering,* 1985, Chapter 6, pp. 1–97.

12. Mil-Hdbk-217F, Section 12.1, December 1991, superceding Mil-Hdbk-217E, Notice 1, January, 1990.

13. Doane, David P., *Exploring Statistics with the IBM P.C.,* (Reading, MA: Addison-Wesley Publishing Company, 1985), pp. 105–133.

14. Chrysler, Ford, and General Motors Supplier Quality Requirements Task Force, *Fundamental Statistical Process Control—Reference Manual,* 1991, pp. 79–82.

15. Chrysler, Ford, and General Motors Supplier Quality Requirements Task Force, *Quality System Requirements QS-9000, 1994.*

16. Adapted from W. G. Cochran, "Sampling Techniques," 2nd Edition, (New York: John Wiley & Sons, 1963). Originally adapted by J. M. Juran, *Quality Control Handbook,* 1979, pp. 24–21 to 24–22.

7

Product Assurance Tools and Processes and Their Application

Production and Continuous Improvement Phase Tools

"Improve constantly and forever every activity in the company, to improve quality and productivity and thus constantly decrease cost."

DEMING'S FIFTH POINT
FRANK PRICE, *Right Every Time*,
ASQ QUALITY PRESS, 1990, P. 89.

Chapter Introduction

This chapter is the last of four chapters involving the strategic use of various quality and reliability tools during each of the product development phases. The purpose of this particular chapter is to discuss the practical application of three techniques utilized in the production and continuous improvement phase of the product development process (refer to the PAP worksheet example, Figure 3.3). These three techniques involve: (1) statistical process control, (2) Pareto charts, and (3) root cause analysis. Incorporation of these tools into the production phase helps to direct the corrective action necessary to resolve the problem and continuously improve the product design and process.

A. Statistical Process Control (SPC)

What and Why the Tool Is Applied

Statistical process control (SPC) is a method for recording, monitoring, and interpreting the degree of nonrandomness in a measured variable or attribute data type characteristic for the purpose of controlling the process. The use of a graphical method or control chart technique for evaluating whether a process is or is not in a "state of statistical control" helps the analyst recognize the period in a process when the variable under investigation exhibits either the tendency for non-randomness variation or being out-of-control. It is then up to the engineering team to determine the root cause for this excessive variation and implement the appropriate corrective action. A check for statistical process control must be made to determine if a capability evaluation can then be made.

When the Tool Is Applied

SPC is primarily used during the production phase of the PDP. The establishment of the type of parameter, the specification limits, and control limits used for a specific chart is usually determined throughout process development and finalized during the verification test stages of the PDP. It is usually through the process control plan that the details of the kind of control chart are first conceived and then further described in the operator instruction sheet.

Where the Tool Is Applied

SPC is applied to any design or process complexity level (component, subsystem, or system level) and part status (new, modified, or carryover). SPC can be implemented at individual process operations or used to help monitor variation at the end-of-line test station.

Who Is Responsible for the Method

Responsibility for implementing and evaluating SPC through the use of graphical control charts would primarily involve technicians and employees at the manufacturing site, with consultation and interpretation from the manufacturing or process, and quality engineers. The product assurance engineer would help to further interpret the control chart and facilitate any root cause analysis effort conducted to reduce the magnitude of variation that might have been observed in the process.

How the Tool Is Implemented

Statistical process control is implemented using either variable or attribute type data and essentially involves trend analysis of the data behavior relative to pre-determined control limits. The SPC would be established through the use of graphical quality control charts set up at strategic positions in the manufacturing line. There are many different kinds of charts for indicating the behavior of a variable or attribute data type characteristic. Some charts for variable type data involve $\overline{X}$ (average) and R (range) charts, s (standard deviation) charts, etc. Some charts for attribute type data may involve p (proportion of defective parts) charts, c (number of events of a given characteristic) charts, np (total number of units in a sample in which an event of a given classification occurs) charts, etc. The procedure for evaluating SPC charts relative to a variable or attribute type data are as follows:

- Variable Chart—$\overline{X}$ and R Chart (see example in Figure 7.1):

1. Select a control characteristic for a specific operation from the process control plan and identify the specification limits.
2. Assure that the measurement system has been defined, calibrated, and undergone a gauge repeatability and reproducibility test prior to the study.

Variable Data Example—X-Bar and R Charts

Given a particular process operation for producing a dimensional clearance between two sections of a part. This clearance has a specification of 34 +/− 6 mm. The manufacturing engineer and quality engineer have agreed to monitor this process through the use of an X-bar and R chart. The development of the chart is highlighted as follows:

1. The dimensional clearance control characteristic was identified through the Process Control Plan.

2. A micrometer was calibrated and recently underwent a gauge R & R study which produced a total GRR of 8 percent. This total variation was small enough to accept the use of the measurement system for the characteristic under investigation.

3. A careful examination of any special causes of variation was conducted prior to the construction of the chart. This process had undergone a process review, which included incoming material evaluation, operator proficiency examination, etc.

4. A subgroup size of 5 was chosen per the sample frequency time interval as specified in the Process Control Plan.

5. A random sample of 100 pieces (20 subgroups of 5 pieces per group) were evaluated per the conditions detailed in the Process Control Plan.

6. The average ($\overline{X}$) and range (R) were calculated for each subgroup using equations #54 and #55, respectively.

7. Each subgroup average and range was plotted on the chart (see example in Figure 7.2).

8. The grand average ($\overline{\overline{X}}$) and average range ($\overline{R}$) was calculated using equations #56 and #57 respectively:

$$\overline{\overline{X}} = \frac{\sum\limits_{x=i}^{n} \overline{X}_i}{n} = \frac{667}{20} = 33.4 \text{ mm}$$

$$\overline{R} = \frac{\sum\limits_{x=i}^{n} R_i}{n} = \frac{116}{20} = 5.8 \text{ mm}$$

9. The grand average was plotted as the centerline for the $\overline{X}$ chart and the average range was plotted as the centerline of the range chart.

10. The control limits for the $\overline{X}$ and R charts were determined as follows:

Using equations #58 and #59 and the constant value found in Table 7a, the upper and lower control limits for the X-bar chart are:

$$UCL_x = \overline{\overline{X}} + (A_2 \times R) = 33.4 + (0.577 \times 5.8) = 36.7 \text{ mm}$$
$$LCL_x = \overline{\overline{X}} - (A_2 \times R) = 33.4 - (0.577 \times 5.8) = 30.1 \text{ mm}$$

Using equations #60 and #61 and the constant value found in Table 7a, the upper and lower control limits for the range chart are:

$$UCL_R = D_4 \times R = 2.114 \times 5.8 = 12.3 \text{ mm}$$
$$LCL_R = D_3 \times R = 0 \times 5.8 = 0 \text{ mm}$$

LCL_R has no lower control limits for subgroup size less than 7.

Continued.

Figure 7.1   Statistical Process Control Examples

11. The average and range charts were analyzed for the degree of statistical control. Upon inspection of both charts there were no points observed outside the specification limits and no points exceeding the process control limits. Further inspection had shown that there were no peculiar trends (a successive number of points on either side of the centerline, an increasing or decreasing movement to the control limit line, hugging the center line, rapid increases or decreases, etc.) in the data, which would indicate the potential for un-natural variation due to equipment, operator, or process methodology.

Attribute Data Example—*P*-Chart

Given a process operation involving a visual inspection of a mechanical subassembly. The quality and manufacturing engineer determined through the Process Control Plan that a *p*-chart should be used to help monitor the particular process step. The development of this chart is highlighted as follows:

1. Visual aids and associated acceptance criteria were chosen for the control technique for the process.
2. The visual inspection system was evaluated using the attribute data measurement system approach to determine repeatability and reproducibility.
3. All variances in the inspection set-up were corrected before the study.
4. The proportion defective was plotted for each subgroup or time interval (see example in Figure 7.3).
5. The average sample size $\bar{n}$ was determined using equation #62:

$$\bar{n} = \frac{n}{z} = \frac{14,091}{19} = 741.6$$

where: n = total number of samples
z = number of subgroups (time intervals)

6. The average fraction defective $\bar{p}$ was determined using equation #63:

$$\bar{p} = \frac{\sum_{x=i}^{n} p_i}{n} = \frac{1030}{14,091} = 0.073$$

where: p_i = individual proportion defectives
n = total number of samples

7. The control limits were calculated using equations #64 and #65:

$$Control\ Limits = \bar{p} +/- 3\sqrt{\frac{\bar{p}(1 - \bar{p})}{\bar{n}}} = 0.073 +/- 0.029$$

8. An examination of the control chart has shown three points outside the limits. The two points which exceeded the upper control limit had indicated a high proportion defective, while the other point exceeded the lower limit indicating a significant reduction in the number of defects. Further observation of the behavior of the data indicates an instable distribution of the proportion defectives in the beginning of the period (significant swings around the centerline) and then a clustering around the centerline followed by a decline in the proportion defective towards the end of the period. This behavior in the data tends to indicate the need for preventive action in the process to help reduce the proportion defective.

Figure 7.1 *Continued.*

TABLE 7a

Table of Constants and Formulas for Control Charts

Subgroup Size	X and R Charts* Chart for Averages (X) Factors for Control Limits	Chart for Ranges (R) Divisors for Estimate of Standard Deviation	Chart for Ranges (R) Factors for Control Limits		X and s Charts* Chart for Averages (X) Factors for Control Limits	Charts for Standard Deviations (s) Divisors for Estimate of Standard Deviation	Charts for Standard Deviations (s) Factors for Control Limits	
n	A_2	d_2	D_3	D_4	A_3	c_4	B_3	B_4
2	1.880	1.128	—	3.267	2.659	0.7979	—	3.267
3	1.023	1.693	—	2.574	1.954	0.8862	—	2.568
4	0.729	2.059	—	2.282	1.628	0.9213	—	2.266
5	0.577	2.326	—	2.114	1.427	0.9400	—	2.089
6	0.483	2.534	—	2.004	1.287	0.9515	0.030	1.970
7	0.419	2.704	0.076	1.924	1.182	0.9594	0.118	1.882
8	0.373	2.847	0.136	1.864	1.099	0.9650	0.185	1.815
9	0.337	2.970	0.184	1.816	1.032	0.9693	0.239	1.761
10	0.308	3.078	0.223	1.777	0.975	0.9727	0.284	1.716
11	0.285	3.173	0.256	1.744	0.927	0.9754	0.321	1.679
12	0.266	3.258	0.283	1.717	0.886	0.9776	0.354	1.646
13	0.249	3.336	0.307	1.693	0.850	0.9794	0.382	1.618
14	0.235	3.407	0.328	1.672	0.817	0.9810	0.406	1.594
15	0.223	3.472	0.347	1.653	0.789	0.9823	0.428	1.572
16	0.212	3.532	0.363	1.637	0.763	0.9835	0.448	1.552
17	0.203	3.588	0.378	1.622	0.739	0.9845	0.466	1.534
18	0.194	3.640	0.391	1.608	0.718	0.9854	0.482	1.518
19	0.187	3.689	0.403	1.597	0.698	0.9862	0.497	1.503
20	0.180	3.735	0.415	1.585	0.680	0.9869	0.510	1.490
21	0.173	3.778	0.425	1.575	0.663	0.9876	0.523	1.477
22	0.167	3.819	0.434	1.566	0.647	0.9882	0.534	1.466
23	0.162	3.858	0.443	1.557	0.633	0.9887	0.545	1.455
24	0.157	3.895	0.451	1.548	0.619	0.9892	0.555	1.445
25	0.153	3.931	0.459	1.541	0.606	0.9896	0.565	1.435

$$UCL_{\bar{X}}, LCL_{\bar{X}} = \bar{\bar{X}} \pm A_2\bar{R}$$
$$UCL_R = D_4\bar{R}$$
$$LCL_R = D_3\bar{R}$$
$$\hat{\sigma} = \bar{R}/d_2$$

$$UCL_{\bar{X}}, LCL_{\bar{X}} = \bar{\bar{X}} \pm A_3 s$$
$$UCL_s = B_4 s$$
$$LCL_s = B_3 s$$
$$\hat{\sigma} = s/c_4$$

*From ASTM publication STP-15D, *Manual on the Presentation of Data and Control Chart Analysis,* 1976; pp. 134–136. Copyright ASTM, 1916 Race Street, Philadelphia, Pennsylvania 19103. Reprinted, with permission.

Source: Chrysler, Ford, and General Motors Supplier Quality Requirement Task Force, *Fundamental Statistical Process Control Reference Manual,* 1991, Appendix E, p. 143. Reprinted with permission.

3. Assure that the process under investigation is void of special causes to variation. Process controls used to reduce the potential of special causes of variation may involve the use of production tools and processes, speed of production, operator understanding and qualification, incoming material sample grouping (homogeneous batch or lot), etc.

4. Determine the subgroup size for the samples taken for a process. For a short-term study, the subgroup size should consist of five or more consecutive parts. A subgroup sample size larger or smaller than five is possible given the nature of the process and part design. The subgroup size should remain constant throughout the study. For a long-term study, the subgroup size should be based on the judgement of the analyst.

5. For a short-term study, a total sample of at least 50 consecutively built samples (10 subgroups of five parts per subgroup) should be randomly chosen for a specific interval of time (per shift, day, etc.). For a long-term study, a total sample of at least 100 consecutively built samples (20 subgroups of at least five parts per subgroup) should be randomly chosen for a longer period of time (per week, etc.). The process control study should consider short and long intervals of time.

6. Calculate the average $\overline{X}$ and range R:

$$\overline{X} = \frac{\sum\limits_{x=i}^{n} x_i}{n} \tag{54}$$

where: $\overline{X}$ = average of samples
x_i = individual samples within the subgroup
n = subgroup sample size

$$R = x_{\text{Highest}} - x_{\text{Lowest}} \tag{55}$$

7. Plot the subgroup averages and ranges on the $\overline{X}$ and R chart as displayed (see example in Figure 7.2).

8. Calculate the grand average, $\overline{\overline{X}}$, and average range, $\overline{R}$, accordingly:

$$\overline{\overline{X}} = \frac{\sum\limits_{x=i}^{k} \overline{X}_i}{k} \tag{56}$$

where: $\overline{\overline{X}}$ = grand average (process average)
$\overline{X}_i$ = subgroup average
k = number of subgroups

$$\overline{R} = \frac{\sum\limits_{x=i}^{k} R_i}{k} \tag{57}$$

where: $\overline{R}$ = average of ranges
R_i = subgroup ranges
k = number of subgroups

9. Plot the grand average as the centerline on the $\overline{X}$ chart and the average range as the centerline on the range chart.

10. Determine the control limits for the $\overline{X}$ and R charts.[1] Refer to Table 7a for identification of the proper constants, relative to subgroup size.

$\overline{X}$ chart control limits:

$$UCL_{\bar{X}} = \overline{\overline{X}} + \left(A_2 \times \overline{R}\right) \tag{58}$$

$$LCL_{\bar{X}} = \overline{\overline{X}} + \left(A_2 \times \overline{R}\right) \tag{59}$$

Data and Control Charts for Performance Study
X̄ and R Charts

Subgroup Number	1	2	3	4	5	6	7	8	9	10	11	12	13	14	15	16	17	18	19	20
Data from Each Subgroup (mm)	38	31	30	30	32	33	33	33	33	30	35	36	30	36	32	28	33	30	35	33
	35	31	30	33	34	38	34	33	36	35	37	38	32	35	35	31	32	31	37	35
	34	34	32	33	37	31	38	36	35	37	35	39	37	40	33	35	34	35	28	35
	33	31	30	32	37	33	30	35	34	32	35	34	30	34	34	31	33	30	27	38
	30	31	32	37	35	33	30	36	31	29	37	30	31	30	35	33	33	30	32	37
Total	170	158	154	165	175	168	165	173	169	163	179	177	160	175	169	158	165	156	159	178
Average ($\bar{X}$)	34.0	31.6	30.8	33.0	35.0	33.6	33.0	34.6	33.8	32.6	35.8	35.4	32.0	35.0	33.8	31.6	33.0	31.2	31.8	35.6
Range (R)	8	3	2	7	5	7	8	3	5	8	2	9	7	10	3	7	2	5	10	5

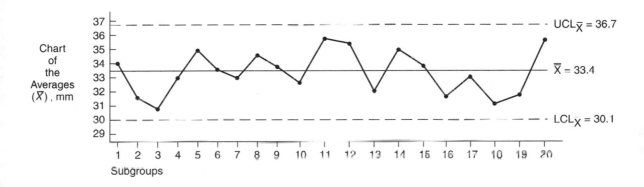

Chart of the Averages ($\bar{X}$), mm

$UCL_{\bar{X}} = 36.7$

$\bar{\bar{X}} = 33.4$

$LCL_{\bar{X}} = 30.1$

Subgroups

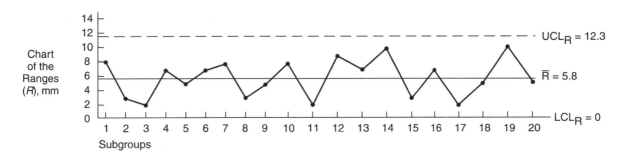

Chart of the Ranges (R), mm

$UCL_R = 12.3$

$\bar{R} = 5.8$

$LCL_R = 0$

Subgroups

Source: General Motors Supplier Development Administration, *General Motors Statistical Process Control Manual (GM-1693),* 1989, p. 4–11. Reprinted with permission.

Figure 7.2 Statistical Process Control Example—$\bar{X}$ and R Charts

where: $UCL_{\bar{X}}$ = upper control limit for $\bar{X}$ chart

$\quad\quad\quad LCL_{\bar{X}}$ = lower control limit for $\bar{X}$ chart

$\quad\quad\quad A_2$ = a constant depending on subgroup size

$\quad\quad\quad \bar{\bar{X}}$ = grand average

$\quad\quad\quad \bar{R}$ = average of the ranges

R chart control limits:

$$UCL_R = D_4 \times \bar{R} \tag{60}$$

$$LCL_R = D_3 \times \bar{R} \tag{61}$$

where: UCL_R = upper control limit for range chart

$\quad\quad\quad LCL_R$ = lower control limit for range chart

$\quad\quad\quad D_4, D_3$ = a constant depending on subgroup size

$\quad\quad\quad \bar{R}$ = average of the ranges

11. Analyze and interpret the chart for the degree of statistical control. In order for a process to be in control, there should be no data points outside the limits or no indication of any unusual patterns which may indicate nonrandomness. In determining trends or patterns leading to variation, one should look for runs (successive points all on one side of the centerline), clustering around the centerline, cyclic patterns, etc.

    Some causes of variation which affect the behavior in an $\bar{X}$ chart might be: change in material, worker, machine, procedure, inspection equipment. If there is a pattern illustrated in the range chart then some cause affecting it may be: change in materials, methods, or worker. Both charts need to be evaluated concurrently in order to make an informed decision on what corrective action may be necessary.

• Attribute Chart—P-Chart (see example in Figure 7.1):

  1–3. Repeat the same steps as for variable control charts. The characteristic that will be monitored will be an attribute data type and will represent the number of defects. The sampling procedure for attribute type data will be based on the number of defects determined for a specified time interval. This time interval will represent a subgroup. The total number of subgroups used for a short- or long-range study will be up to the discretion of the engineering team.

  4. Plot the subgroup proportion defectives on the p-chart as displayed (see example in Figure 7.3).

  5. Calculate the average sample size and the average proportion defective for the number of samples defective:

$$\bar{n} = \frac{n}{k} \tag{62}$$

where: $\bar{n}$ = average sample size

$\quad\quad\quad n$ = total number of samples

$\quad\quad\quad k$ = total number of subgroups or time intervals

$$\bar{P} = \frac{\sum\limits_{i=1}^{k} n_i p_i}{n} \tag{63}$$

where: $\bar{p}$ = average proportion defective

$\quad\quad\quad n_i p_i$ = number of nonconforming samples per subgroup

$\quad\quad\quad n_i$ = subgroup sample size

$\quad\quad\quad p_i$ = subgroup proportion defective

$\quad\quad\quad n$ = total number of samples

$\quad\quad\quad k$ = total number of subgroups or time intervals

Sample Period	No. of Inspected Samples	Defective Samples	Fraction Defective (P_i)
1	724	48	0.067
2	763	83	0.109
3	748	70	0.094
4	748	85	0.114
5	724	45	0.062
6	727	56	0.077
7	726	48	0.066
8	719	67	0.093
9	759	37	0.049
10	745	52	0.070
11	736	47	0.064
12	739	50	0.068
13	723	47	0.065
14	748	57	0.076
15	770	51	0.066
16	756	71	0.094
17	719	53	0.074
18	757	34	0.045
19	760	29	0.038
Totals	14,091	1030	
Averages	741.6	54.2	0.073

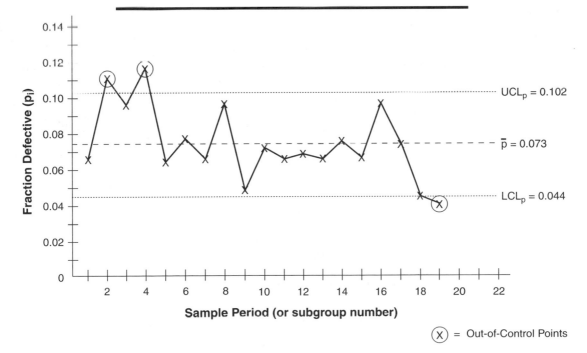

Source: J. M. Juran and Frank M. Gryna Jr., *Quality Planning and Analysis: From Product Development through Use,* Second Edition, (New York: McGraw-Hill, 1980), p. 339. Adapted by permission.

Figure 7.3 Statistical Process Control Example—P-Chart

6. Plot the average proportion defective as the centerline on the *p*-chart.
7. Determine the control limits for the *p*-chart as follows:[2]

$$UCL_p = \bar{p} + 3\sqrt{\frac{\bar{p}(1-\bar{p})}{\bar{n}}} \qquad (64)$$

$$LCL_p = \bar{p} - 3\sqrt{\frac{\bar{p}(1-\bar{p})}{\bar{n}}} \qquad (65)$$

where: UCL_p = upper control limit
LCL_p = lower control limit
$\bar{p}$ = average proportion defective
$\bar{n}$ = average sample size

8. Analyze and interpret the chart for the degree of statistical control. In order for a process to be in control, there should be no data points outside the limits or a display of any unusual patterns which may indicate nonrandomness. In determining trends or patterns leading to variation, one should look for runs (successive points all on one side of centerline), clustering around the centerline, cyclic patterns, sudden change in level, cumulative trend, etc.

The causes of variation affecting the *p*-chart are similar to the $\bar{X}$ and *R* charts and involve material, worker, machine, procedure, inspection equipment. The patterns to look for, as was for the variable charts, are: lack of stability, cumulative trends, cyclical, and sudden change in level. The *p*-chart needs to be evaluated frequently in order to make an informed decision on what corrective action may be necessary to improve the process.

Benefits of the Tool

The control chart is a very useful and powerful statistical tool for monitoring process control parameters. It is a tool which helps to indicate nonrandomness or out-of-control situations for a process. Once these charts have proved their effectiveness and the root causes of variation are identified and corrected, they should be removed and substituted with error- or mistake-proofing features in the product design or process. Some of the immediate benefits of process control charts are as follows:

1. Provides an indication of product design/process stability by displaying data trends for a parameter over time.
2. Allows the analyst the ability to first identify significant process variation or an out-of-control condition and provide appropriate preventive measures.
3. Functions as a simple and practical detection tool for the operator, mechanic, or technician to use and interpret. Statistical process control gives the operator an opportunity to regulate his or her process.
4. Provides useful information to engineering personnel and the product assurance engineer for the root cause of a particular problem.

B. Pareto Chart Analysis

What and Why the Tool Is Applied

Pareto chart analysis is a technique for the evaluation of frequent events by category (Pareto chart principle) and the relationship of the events from one chart to the other. This method involves the construction of frequency histogram graphs and the general comparison of events of one chart to another in order to identify possible relationships leading to root cause analysis. The Pareto chart approach is a very simple means of displaying and initiating cause-and-effect determination.

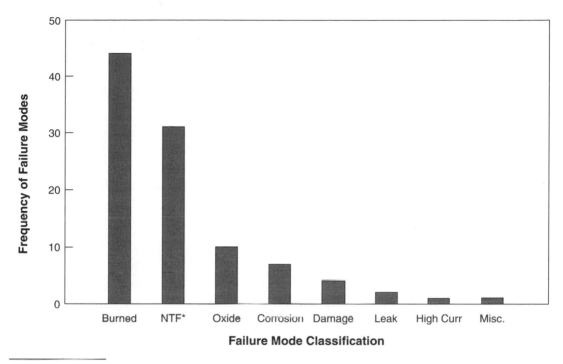

Note: NTF = No Trouble Found

Figure 7.4   Pareto Chart Example—Total Field Returns of a Product

When the Tool Is Applied

The application of Pareto chart analysis can occur throughout the product development cycle and helps to complement the cause-and-effect process of root cause analysis, which also can be applied anywhere during product development. It is commonly used during production when there is sufficient sample size of product in which to examine proportion of event occurrence.

Where the Tool Is Applied

Pareto chart analysis can be applied on any design or process complexity level (component, subsystem, or system level) and part status (new, modified, or carryover).

Who Is Responsible for the Method

The responsibility for constructing and interpreting the Pareto charts is open to any member of the engineering team who may feel that the event is better represented over other events in a simplified manner. This method can be easily applied by any member of the team to represent an event during any phase of the product development process. This tool is a popular means of summarizing and interpreting events for management.

How the Tool Is Implemented

The construction and analysis technique for Pareto charts basically involves developing a series of related frequency distribution graphs and inspecting the relationship of the most frequent events within the chart (see example in Figure 7.4). The basic process is represented as follows:

1. Determine the type of categories for the events to be compared with one another (x-axis).
2. Count the number per category and illustrate this frequency as a bar per the category (y-axis).

3. Compare the bars per category and note the relationship between categories. Identify the most or least frequent event categories.
4. Construct frequency histograms relating other components for the same categories previously investigated and plot the frequency distribution.
5. Compare the amplitudes or frequency of the events in other graphs and look for relationships.

Benefits of the Tool

Pareto charts provide a quick and easy way to compare event categories with another or to related charts. A summary of the benefits from Pareto charts include the following:

1. Provides important information on the frequency of a desired event which may be related to the root cause of a problem under investigation.
2. Develops basic cause-and-effect inferences, which help in the development of fishbone diagrams, etc.
3. Provides an easy graphical means of displaying frequency distribution data and developing basic conclusions on the data.

C. Root Cause Analysis (RCA)

What and Why the Tool Is Applied

Root cause analysis (RCA) is a process of evaluating a problem to the extent of identifying the failure mechanism or cause which produced the original problem. The root cause process involves a logical sequence of determining the problem statement, developing potential causes, evaluating the causes, isolating the main cause, and validating the main cause. The process can be thought of as a logical sequence of addressing the what, why, when, where, and who, in order to address how the cause occurred. This tool is applied to help understand the details behind the problems which have been experienced in order to determine the appropriate preventive action.

When the Tool Is Applied

Root cause analysis is usually applied during the production phase of the PDP, but can be more effective if applied during the product design/process development and validation phases of the PDP. During the production phase, application of root cause analysis is common since corrective action on failures exhibited in the production demands immediate attention. However, in order to reduce the risk of directing the organization to the wrong corrective action it is advisable to incorporate RCA during the development phase. Defining the failure mechanism and quickly removing it from a design is of significant advantage to the engineering team during the design development phase, since the cost of the engineering change in the development phase is less than the cost incurred during the production phase.

Where the Tool Is Applied

RCA is applied to any design or process complexity level (component, subsystem, or system level) and part status (new, modified, or carryover).

Who Is Responsible for the Method

Participants in root cause analysis would involve many representatives of the engineering organization (design, manufacturing, service, etc.) and include the quality, reliability, or product assurance personnel. The chairman may be an engineering representative of an area from which

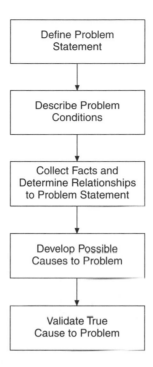

Figure 7.5   Root Cause Analysis Process

the problem initially evolved, but may be chaired by the product assurance specialist. The PA specialist would help the group by contributing background material and consultation toward experimentation and analysis, and direct the logical root cause analysis sequence.

How the Tool Is Implemented

The execution of root cause analysis primarily involves a step-by-step review and evaluation of the problem to possible causes. The implementation of each step in the process (see Figure 7.5) involves the collective effort of the group to address each objective with supporting data, analysis, and interpretation which would lead to the cause. A brief review of the process is as follows (see example in Figures 7.6 and 7.7):

1. Define the problem statement and address the conditions related to the problem. State the elements which may or may not be related to the specific problem statement.
2. Describe and define possible causes to the specified problem. Address each possible cause with background material that may support or disprove the relationship to the problem. Obtain this information from the FMEA, FTA, engineering failure analysis, experimental test results, simulation studies, previous test results, etc. A development of possible hypotheses, which may be tested quantitatively or qualitatively, is needed to explore cause and effect.
3. Evaluate the list of possible causes through the use of statistical tools or engineering evaluation and isolate the most likely candidate for the true cause of the problem. The evaluation approaches used can involve quantitative statistical approaches such as test of hypotheses and test analysis techniques. If the data is qualitative in nature, then the use of decision-making techniques (with or without probabilistic data) may be incorporated to

Problem Statement:

Sound-producing musical device not delivering any discernible sound.

Problem Conditions and Facts:

What: Model T, Brand X sound-producing device not producing sound.
Where: Region #4 out of five possible regions in marketing environment.
When: Two weeks ago, consistently.
Extent: 300 parts per 10,000 or three parts per every 100 produced.
        All devices exhibit no audible sound.
        Devices are failing at an increasing rate.

Possible Problem Contributors (Causes):

1. Shorted positive terminal inside device
2. Excessive oxide build-up on electrical contacts
3. Loss of stand-alone controller electrical ground
4. Open circuit in sound-producing device system
5. Extreme temperature or humidity reaction to components of sound-producing device
6. Mishandling of device upon shipping or assembly
7. Switch button sticking

Evaluate the Causes (see Figure 7.7):

Statistical Tool #1: A cause and effect diagram was constructed from previous failure mode experiences in the field and from the current FMEA.
Statistical Tool #2: A Pareto Chart was then constructed on the distribution of failure modes recently experienced. This Pareto Chart was compared to an older chart for a previous design. A potential for inoperative sound-producing devices based on oxidized contacts was experienced in the past, but to an insignificant level. An inspection of the current Pareto Chart based on parts produced in the three-month period were definitely related to an oxidized contact condition.
Statistical Tool #3: A designed experiment was conducted to investigate the factors that may contribute to oxidized contacts. Results indicated that the handling techniques, coupled with the material composition of a particular lot of contact material had a significant influence on the generation of corroded contacts. Further investigation had uncovered that the mishandling of the contacts due to moisture from operator hands reacted with a contact material that was not properly heat treated to remove all the oxidants present in the alloy fabrication process.

Confirmation of the Identified Cause:

A confirmation run was performed using the same conditions as defined in the original designed experiment. Results of the second test confirmed the conclusions reached in the first experiment.

Corrective Action:

Corrective action was conducted to enhance the process control of the operator handling of the material during assembly. Gloves were require at this particular operation. In addition, the heat treatment process of the alloy was carefully reviewed and resulted in the identification of impurities entering the annealing phase and thus producing a reactive alloy to oxygen.

Figure 7.6   Root Cause Analysis Example—Sounding Device Failure

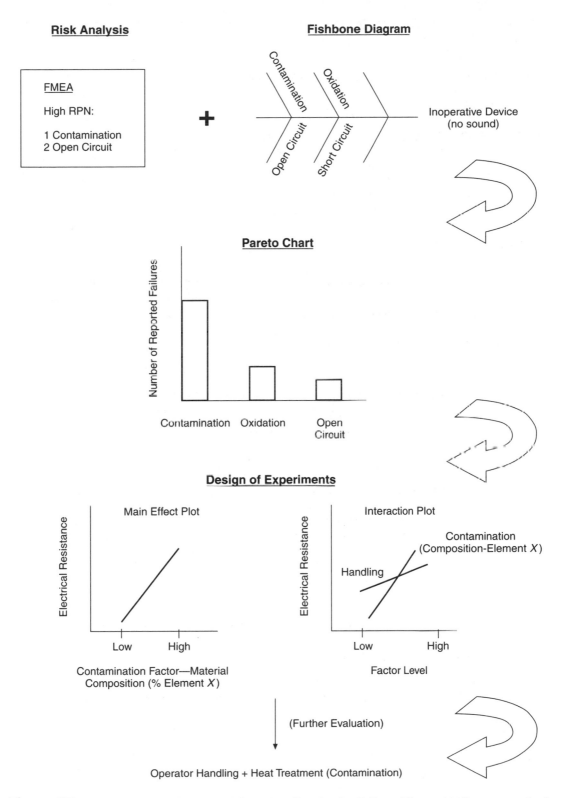

Figure 7.7 Root Cause Analysis Example—Sounding Device Failure (Illustrative Representation)

identify a true cause. In either case, a formal (decision tree or payoff table) or informal (comparison, etc.) decision-making technique can be used to narrow the possible causes to the true cause.

4. Validate the selection of the true cause through performing tests, experiments, process demonstrations, etc., which would help to reproduce the problem. Repeat as many times as necessary in order to build confidence under variations of environmental conditions.

Benefits of the Tool

Root cause analysis provides a common sense approach to evaluating a problem, asking the right kinds of questions, and directing the thought for establishing and locating possible causes to the problem. It is a tool to be used for supporting the determination of failure mechanisms of product designs and processes. Highlights to the benefits of root cause analysis are as follows:

1. Provides a logical thought process for directing the identification and verification of causes to specific problems.
2. Helps formalize the means to organizing and accomplishing identification of failure mechanisms. Root cause analysis helps justify these failure mechanisms for the FMEA, which is the central document for risk assessment of product designs and processes.
3. Presents a simplified but adaptable means for determining possible causes and evaluating those causes.
4. Adapts during any phase of the product development process. Root cause analysis is most beneficial when coupled with failure analysis after testing or preliminary production process verification.
5. Involves many other areas of engineering in the determination of a root cause. Root cause analysis brings groups together for the common goal of problem resolution.

NOTES

1. Chrysler, Ford, and General Motors Supplier Quality Requirements Task Force, *Fundamental Statistical Process Control—Reference Manual,* 1991, pp. 29–57.
2. ———— pp. 89–105.

8

Product Assurance Training

"Institute a vigorous program of education and re-training. New skills are required for changes in techniques, materials, and services."

DEMING'S THIRTEENTH POINT
FRANK PRICE, *Right Every Time*
ASQ QUALITY PRESS, 1990, P. 105

Chapter Introduction

The third major element in practical product assurance management is just-in-time product assurance training. The objective of this chapter is to discuss the training necessary to prepare the quality/reliability professional for implementation of product assurance principles in an organization. Topics include: (1) what kind of training (short and long term), (2) reasons for training, (3) when training is performed, (4) where training is applied, (5) who does the training, and (6) how training is performed. This chapter will help guide management to proper development of the product assurance organization.

The training of product assurance principles is an essential part of product assurance management. These principles are the essential tools for product assurance personnel to use in assisting the development of a product design or process. Just as these principles are important to product assurance engineers, it is important that the engineering team understand the basic idea and use of these techniques and tools so that they may benefit from the results, which aid in proper product design and process development. This chapter will discuss the kind of product assurance training, reasons for the training, when the training is to be given (the JIT concept), where it is used, who receives the training, and how the training should be given.

What Kind of Training

Product assurance training covers many different areas of product development and is adaptable for evaluating and improving product designs and processes. These tools and concepts can be applied to any kind of product design (electrical, mechanical, electromechanical, chemical, etc.) and process (batch, continuous flow, etc.). A familiarization of the product design and process prior to the application of these various product assurance concepts and techniques is essential in order for the results of the product assurance activity to add significant value to product design/ process development.

The type of product assurance training can be divided into short- versus long-term tools and concepts. In other words, some concepts should be understood and applied just-in-time (short term) during product development, while other tools should be considered in the near future as a part of the normal product development process. A summary of these tools is displayed in Figure 8.1.

The short-term product assurance concepts and techniques which are applied during concept and design development, involve quality function deployment, risk analysis, basic quality and reliability principles and evaluation, test design, sample planning, and simulation. These techniques are requirements for the product assurance engineer. However, as new techniques evolve in these areas, the product assurance engineer must learn or be aware of these improvements and communicate them to the engineering team. Such improvements or enhancements to these design tools might be standard FMEA format and analysis technique, simplified reliability or quality analysis approaches for practical engineering application (test data analysis), reliability test plan development and philosophy, etc.

The short-term product assurance concepts and techniques, which are applied during the process development phase, involve process control plan development, measurement systems analysis, process performance studies, and preventive maintenance. Just as they were during design development, these process development techniques are basic requirements for the product assurance engineer. He or she must be aware of new and better ways of improving the use of these techniques for the engineering team. Such enhancements to these process techniques might involve standard process control plan development, adaptable measurement system analysis for attribute and variable gauge measurement systems and conditions, tool and equipment studies and guidelines, simple maintainability evaluation techniques, and scheduled PM plan development.

The long-term product assurance concepts and techniques, which are considered for use in near future programs, involve product design and process simulation activities, accelerated testing concepts and programs, and improved correlation of field to test data. These areas are important subjects for product assurance development and training. Product assurance personnel should strive to clearly understand the techniques and apply these concepts into the product development activities of a traditional PDP. These tools and concepts will facilitate understanding of product design/process behavior, decrease development time, improve quality and reliability, and relate techniques performed in the laboratory to field results.

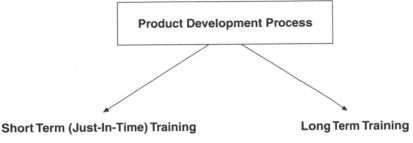

Figure 8.1 Types of Short- and Long-Term Training Material

Other concepts and tools which a product assurance engineer should be trained and proficient in if he or she is to be the facilitator and consultant for product design and process improvement involve:

1. Expertise in product assurance planning and the understanding of where and when specific Q and R tools should be applied.
2. Knowledge and comfort with the specific product design and process operation. The ability to communicate with the design and process engineers on specific details of the product's functionality and construction is important.
3. Familiarization with the access and use of information databases containing data on product warranty, specifications, design guidelines, field history, etc.
4. Awareness and practical use of the latest statistical and engineering simulation software tools. The PA engineer should be comfortable with the use of various quality and reliability software tools (Statgraphics, Minitab, Weibull Smith, etc.). He or she should also be comfortable and proficient in the use and interpretation of various simulation tools such as variation simulation analysis, finite element analysis (basic use and interpretation of output), etc. Combining the results of engineering simulation and statistical analysis helps provide the product engineer with support data necessary for directing design/process development. These software packages become the essential tools for the PA engineer to use in problem solving, consultation, and training.
5. Understanding of all engineering milestones of the product development process and the relationship of product assurance activities (PAP).
6. Practical application and interpretation of quality cost analysis and management. This should include the understanding of life-cycle cost analysis.
7. Specialized training on the details of various Q and R tools such as QFD, test sample planning, quality and reliability analysis, etc.
8. Proficiency in the application of risk analysis (FMEA, FTA, etc.).

9. Familiarization and basic knowledge of software reliability. This area of reliability engineering is a new frontier and would require a comprehension of basic software development, software test planning and analysis, and the implementation of software assurance planning (SAP).

10. Ability to facilitate and actively participate in the design review process. The PA engineer should be very familiar with the appropriate topics to be covered and the appropriate questions to address concerning product design and process development.

Reasons for Training

Product assurance training is an essential part of the key elements to practical product assurance management. By updating PA personnel on the latest developments in the quality and reliability field and training the engineering community on these concepts, the product design and process benefits through the improvements early in product development. The just-in-time concept of training product assurance personnel and the engineering community, including management, is necessary in order to introduce new approaches or reinforce the practical application of particular PA tools that help to build reliability and quality directly into the product design and process. Without any proper training in product assurance principles, the engineering community loses insight on how to better engineer their product design and process effectively and eliminate costly failures at the production facility and in the field. Therefore, it is important to understand the attributes of product assurance training and exactly how they directly affect product design and manufacture.

When Training Is Performed

Product assurance training is performed on a just-in-time basis during the product development process for both product assurance personnel and the engineering staff (including management). The training of product assurance would precede training of the engineering community in order for the PA organization to evaluate and apply the merits of any learned techniques that may be beneficial to product design/process development and validation. The training of the engineering staff would be strategically positioned several weeks to a month in advance of the proposed usage with the tool on actual product/process development. This time is important in order for the user to become more familiar and comfortable about the kind of tool and how it may be used to gain the most benefit.

Where Training Is Applied

Product assurance training can be applied to facilitate development and analysis of either hardware or software product designs or manufacturing processes. It is applicable to the development of components, subsystems, and system level product designs and processes. The kind and amount of training for either hardware or software product design/process would be applied at a very practical level in order to gain immediate benefit

Who Does the Training

Responsibility for product assurance training of the engineering community rests with members of the PA organization and/or contracted individuals either from the organization or from the outside. The amount of detail needed for simple and practical application of a specific Q and R tool or process will direct the amount of preparation and effort necessary to communicate the tool to the engineering community in a just-in-time manner. For engineering management personnel, a simple overview of the techniques and their relationship to the support of the PDP is adequate. However, for the engineering staff, a more detailed overview of the techniques and how they work from a very practical standpoint is necessary. The length of time for effective instruction and comprehension from the audience may be one to two hours maximum per Q and R tool or process. Qualifications for the trainer and trainee of product assurance principles must be carefully considered if a successful training program is to be achieved (see Figure 8.2).

Trainer Qualifications

- Certified Quality and Reliability Engineer
- Minimum 7–10 Years Experience in the Application of Product Assurance
- An Accredited Bachelor's/Master's Degree in Engineering
- Teaching Experience (minimum of five years)
- Other Degrees in Management or Business Desired
- Familiarization of Product Designs and Processes
- Good Interpersonal Skills—Communication Teamwork, Ambition, etc.

Trainee Qualifications

- Management, Product Engineering, Supplier Quality Attendees
- Engineering or Other Technical Degree Requirement
- Familiarization of Product Development Process of Organization
- Willingness to Learn and Participate

Figure 8.2 Qualifications for Product Assurance Trainers and Trainees

How Training Is Performed

The strategy for training either the product assurance engineer or the engineering community must consider a logical sequence in order for it to be effective. For product assurance engineers, the training received by professionals in the field should follow a brief specialized instruction in simulation and its relationship to traditional quality and reliability tools and processes. This focus helps the product assurance engineer relate engineering tools to the statistical approaches in order to better predict and advise the engineering community on product design and process risks. For engineering personnel, the path of training should include how the inputs and outputs from various Q and R tools and processes relate with one another in fulfilling an engineering objective. Therefore, the instruction strategy will involve a brief explanation of how the tools interrelate and then a session-by-session overview of the mechanics of the individual techniques. Some of these techniques may be combined to reinforce the logic behind the analysis or decision-making process.

Training the engineering community in product assurance principles should consider several key areas. The following is a list of useful guidelines in the training process:

1. Train the engineering personnel first and receive total commitment from management before embarking on the just-in-time training strategy for the engineering community.
2. Obtain management attendance and participation on as many of the training sessions for engineering as possible. Ensure management attendance and involvement in the overview training of the product assurance process tool(s).
3. Emphasize team commitment between design, manufacturing, quality and reliability, and management. All parties must have familiarization of work to accomplish tool objectives and properly communicate I/Os with one another. Gain management support in every way.
4. Demonstrate the interaction between the use of Q and R tools and processes within various phases of the product development cycle. Use actual test cases and experiences to support the methodology.

5. Use simplified mathematics, easy to visualize diagrams and graphs, and common sense explanations to cover the material.

6. Utilize any computer software tools to aid in the Q and R analysis process. Focus on the problem-solving approach to the specific tool and emphasize the interpretation of the results.

7. Explain the statistical significance or confidence behind various engineering tools used in a specific technique or process. State assumptions and conditions used in the techniques.

8. Demonstrate the use of the techniques with relevant examples.

9. Stress when/where these tools are applied in order to deliver the maximum value in helping solve a problem or initiating design/process improvement.

10. Bring in actual work for classroom discussion and debate problem-solving approach. Reverse roles in the problem-solving approach and interpretation.

9

Cost of Quality Management

"The more complicated products become, the more reliable components must be if costs are to be held down. Poor work affects expenses all along the line—in scrappage, repairs, larger inventories to provide a cushion against defective parts, higher warranty costs, and eventually lost reputation and sales."

JEREMY MAIN
"THE BATTLE FOR QUALITY BEGINS,"
Fortune, DEC. 29, 1980
PP. 28–33

Chapter Introduction

The fourth and last major element of practical product assurance management is cost of quality management. The primary purpose of this chapter is to help the reader understand the basic approaches to cost of quality management: (1) how to evaluate the total cost and its cost components, and (2) life-cycle cost analysis during the product development process. Both techniques are very helpful in guiding management toward strategic planning of the product assurance effort and a measurement of quality initiatives.

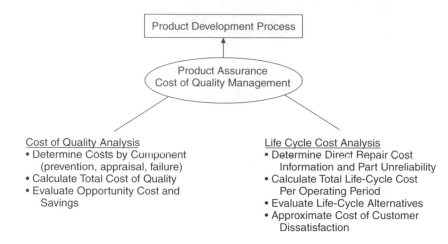

Figure 9.1   Product Assurance Cost of Quality Management Techniques

A. Cost of Quality Analysis

The cost of quality in any organization, whether a manufacturing or nonmanufacturing business, can be viewed as a means of identifying where quality problems are located and how much they may cost the organization during the product development process. These costs indicate to management where effort is needed to adjust and reduce the total cost of quality. It should be understood that quality costs do not solve problems but are key indicators of how quality can be improved and costs reduced in the development of a particular product. Two quality cost analysis techniques are involved in cost of quality management. These techniques are the traditional cost of quality analysis and life-cycle cost analysis (see Figure 9.1).

Cost of quality analysis can be performed by identifying and comparing the magnitude of costs incurred in specific cost categories. These cost categories can be classified as either input or output cost elements (see Figure 9.2). The input cost elements contain prevention and appraisal related expenses. Prevention related expenses involve those costs incurred which, either in the short or long term, help eliminate or avoid future expenses incurred due to failures or variation of the design and process. Appraisal related expenses involve those costs incurred in inspecting or testing the product for signs of failure or special variations in the design or process. On the other hand, there are output cost elements of a design or manufacturing process which contain internal and external failure costs. The internal failure costs refer to defects found and repairs made during the manufacture of the product, while external failure costs refer to failures in the field or complaints received by the customer. These four categories of quality cost and their comparison constitute the basis behind quality cost analysis.

The implementation of quality cost management can be described through evaluating quality cost components and predicting the life-cycle cost of the product design/process during the product development cycle. The evaluation of quality cost components involves: (1) collecting the quality cost information, (2) analyzing trends and patterns, 3) initiating corrective actions based on these results, and (4) reevaluating these results compared to previous situations or over time. This procedure can be reviewed in Figure 9.3. The goal in quality cost analysis is to analyze where to implement quality effort in order to shift costs from failure-related expenses to the prevention-related expenses in the short term while minimizing the total cost of quality in the long term. This total cost of quality is the sum of all the quality cost components. A lower total cost of quality equates to higher profit margins and a higher customer satisfaction level due to improved quality and reliability, which is brought on by preventive-oriented activities.

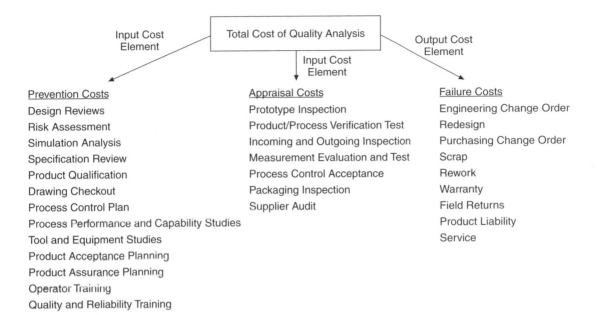

Figure 9.2   Product Assurance Quality Cost Elements and Their Components

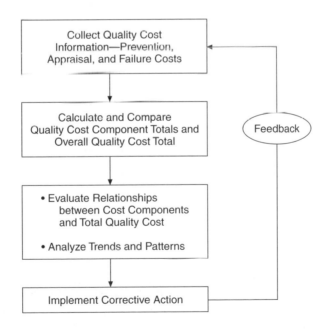

Figure 9.3   Cost of Quality Analysis Procedure

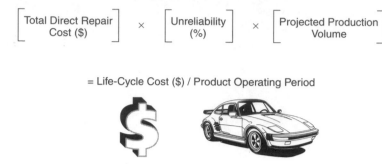

$$\left[\begin{array}{c}\text{Total Direct Repair} \\ \text{Cost (\$)}\end{array}\right] \times \left[\begin{array}{c}\text{Unreliability} \\ \text{(\%)}\end{array}\right] \times \left[\begin{array}{c}\text{Projected Production} \\ \text{Volume}\end{array}\right]$$

= Life-Cycle Cost (\$) / Product Operating Period

Figure 9.4 A Simplified Life-Cycle Cost Formulation

A summary of the key elements of quality cost analysis are as follows:

1. Directs management as to where to apply the appropriate quality effort in minimizing the total cost of quality.
2. Provides a process of measuring the output of quality effort and comparing the costs for various alternatives.
3. Allows management to establish targets for quality and reliability, cost reduction, and plan actions to meet these targets.
4. Enables managers to recognize the magnitude of the quality and reliability opportunity.

B. Life-Cycle Cost Analysis

Life-cycle cost (LCC) analysis is another activity that is pursued by product assurance management to help predict the total life-cycle cost of a product design.[1] It is a methodology which combines reliability engineering and cost accounting techniques to facilitate evaluation of short- and long-term product effectiveness early in the development cycle. The LCC technique essentially involves combining unreliability data with apportioned direct repair costs to produce a life-cycle cost estimation for some particular operating period. Depending on the design level at the time of the LCC analysis, unreliability data may be obtained from warranty, full-stress reliability prediction, or reliability test sources. Concurrent with obtaining this information, repair cost data are collected from various business departments. All of these data are input into a LCC analysis process that translates the information into a liability cost of a design, another factor in determining product design feasibility early in a product development cycle.

The determination of the total life-cycle cost can be simplified to a mathematical product of the total direct repair cost, the unreliability, and the projected production volume, and then expressed per a specific operating period (see Figure 9.4). This life-cycle cost evaluation process can be conducted through: (1) input of cost information from purchasing, finance, service, etc., (2) development of a direct repair cost matrix using the cost information, (3) input of reliability information and cost data from the repair cost matrix to the development of a life-cycle cost breakdown profile, and (4) presentation of a cost impact summary. A review of this analysis sequence is found in Figure 9.5. The resulting information from the cost impact summary matrix helps compare design/process alternatives, predict possible cost/benefits, and facilitate the managerial decision process during development (see example in Figures 9.6a, 9.6b, and 9.6c). To some degree, the total company loss and life-cycle cost can be compared to understand the margin of the cost of customer dissatisfaction as portrayed in Figure 9.7.

The LCC analysis process can be used in other ways to augment cost of quality management. This LCC approach can be used to determine the impact of a particular design's reliability on sales

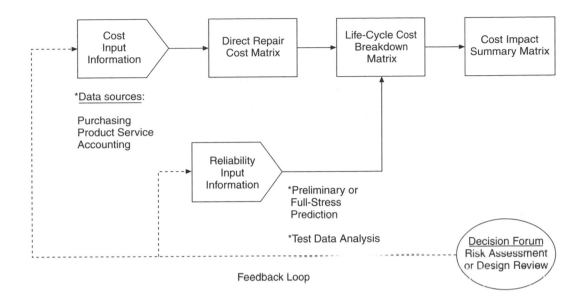

Source. John Bieda, "A Product-Cycle Cost-Analysis Process and Its Application to the Automotive Environment," *1992 Proceedings—Annual Reliability and Maintainability Symposium.* (New York: © IEEE, 1992) p. 423. Adapted by permission.

Figure 9.5 Life-Cycle Cost Analysis Flowchart

	Widget 1		Widget 2	
Cost Category	System Failure	System No Trouble Found	System Failure	System No Trouble Found
Dealer removal and replacement	$18.00	$18.00	$18.00	$18.00
Shipping: Dealer to repair center	—	—	$ 3.00	$ 3.00
Exchange stock cost	—	—	$15.00	15.00
Average part cost: Service center to company	$1.80	$1.80	$3.43	—
Labor cost at repair	—	—	$28.80	$18.80
Company processing	—	—	$ 5.00	$ 5.00
Development cost	—	—	$ 2.65	—
Shipping: Repair center to Dealer	—	—	$ 3.00	$ 3.00
Total Direct Repair Cost	$19.80	$19.80	$78.88	$62.00

Source: John Bieda, "A Product-Cycle Cost-Analysis Process and Its Application to the Automotive Environment," *1992 Proceedings—Annual Reliability and Maintainability Symposium,* (New York: © IEEE, 1992), p. 423. Adapted by permission.

Figure 9.6a Life-Cycle Cost Analysis Example—Direct Repair Cost Matrix

	Widget 1		Widget 2	
Parameter	**System Failure**	**System No Trouble Found**	**System Failure**	**System No Trouble Found**
Incident rate (Failures / 1E+06 Hrs.)	14.94	2.54	5.951	1.01
Number of Incidents (1st Year)	29,287	4,990	11,685	1,985
Number of Incidents (5 Years)	120,998	20,763	48,522	8,266
Direct repair cost	$19.80	$19.80	$78.88	$62.00
Life Cycle Cost (5-year warranty)	$2,395,760	$411,107	$3,827,415	$512,492
Total Life Cycle Cost (Failure and NTF)	$ 2,806,867		$ 4,339,907	
Total Life Cycle Cost (5-year warranty for 4 model years)	$11,227,468		$17,359,628	

Notes: 1. Approximate total system production per year = 455,000 (Average 12 Widgets per system)
     2. Customer dissatisfaction was not included in this analysis.

Source: John Bieda, "A Product-Cycle Cost-Analysis Process and Its Application to the Automotive Environment," *1992 Proceedings—Annual Reliability and Maintainability Symposium,* (New York: © IEEE, 1992), p. 424. Adapted by permission.

Figure 9.6b   Life-Cycle Cost Analysis Example—Cost Breakdown Matrix

	1 Year Volume ($)		4 Year Volume ($)	
Parameter	**Widget 1**	**Widget 2**	**Widget 1**	**Widget 2**
Piece Cost	624,078	1,411,107	2,496,312	5,644,428
Total Life Cycle Cost (5-year warranty period)	2,806,867	4,339,907	11,227,468	17,359,628
Total Cost Impact	3,430,945	5,751,014	13,723,780	23,004,056

Source: John Bieda, "A Product-Cycle Cost-Analysis Process and Its Application to the Automotive Environment," *1992 Proceedings—Annual Reliability and Maintainability Symposium,* (New York: © IEEE, 1992), p. 424. Adapted by permission.

Figure 9.6c   Life-Cycle Cost Analysis Example—Cost Impact Summary Matrix

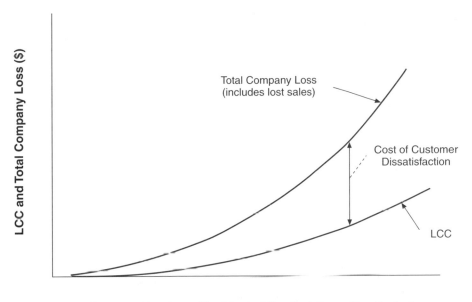

Source: John Bieda, "A Product-Cycle Cost-Analysis Process and Its Application to the Automotive Environment," *1992 Proceedings—Annual Reliability and Maintainability Symposium,* (New York: © IEEE, 1992), p. 424. Adapted by permission.

Figure 9.7   Comparison of Total Company Loss and LCC for a Product Operating Period—Example

potential. Understanding how changes in a design affect total warranty per part will help to minimize customer dissatisfaction and increase a company's profit. Although the direct cost of lost sales due to customer dissatisfaction has not been well documented, it is generally assumed that this cost increases exponentially as customer incidents increase. The life-cycle cost curve is determined by generating an LCC versus number of incidents for a range of mission times. The total company loss curve is the sum of these two.

A sensitivity analysis of the total company loss per part would help to indicate the impact of the reliability of various design alternatives. By observing what kind and how much of a design change results in significant differences in life-cycle cost and total company loss per part, a better understanding can be obtained of those product variables (manufacturing, design, assembly, services, etc.) which appreciably affect consumer behavior for the life of the part.

A summary of the key elements provided from a life-cycle cost analysis are as follows:

1. Product unreliability and direct repair cost data can be combined to help forecast product warranty and evaluate design alternatives.
2. LCC analysis can also be used with target costing techniques to help understand total cost and profit potential for marketing strategy purposes.
3. Cost of customer dissatisfaction may be better understood as the LCC curve is compared to total company loss (including lost sales).
4. LCC analysis promotes cooperation and teamwork between engineering and financial departments.

C. Product Assurance and the Cost of Quality Management

The practical approach to product assurance management is to utilize cost of quality techniques to help benchmark and correlate product improvement activity to actual field behavior during product development. The implementation of quality cost and life-cycle cost analyses helps the product assurance organization demonstrate the effectiveness of its other activities such as PAP, Q and R tool implementation, and training to the successful development of product design and process during the PDP.

The product assurance organization is helpful in the determination of where in the PDP the quality costs are incurred and how to shift the costs toward preventive activities, which in the short and long term help to minimize the total cost of quality. The ability of PA to significantly impact the transfer of failure costs to preventive costs can occur as more effort is spent early in the PDP cycle with product design and process engineering. This is accomplished through understanding risk, simulating product/process performance under a stress environment, and testing more intelligently and efficiently. In addition, product assurance assists in the goal for lower total cost of quality by concentrating its support to product design and process development early in the PDP where the opportunity for product evaluation is more economical, less stressful, and more opportune for design and process analysis, benchmarking, testing, and risk

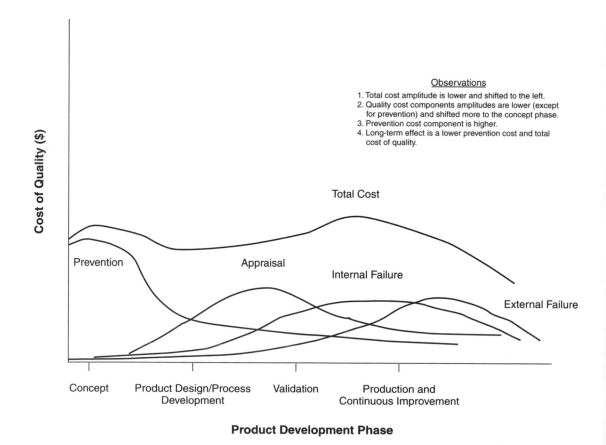

Figure 9.8a   Quality Costs and the Product Development Process (Desired Condition)

evaluation. As engineering effort is expended in the early stages of the PDP, prevention costs will pay for themselves in time and will have the overall effect of lowering the total cost of quality, which is the sum of preventive, appraisal, and failure costs (see Figures 9.8a and 9.8b and the example in Figure 9.9).

The following is a list of product assurance tasks associated with cost of quality management:

1. Cost of quality management helps to compare and contrast the PA organization and the engineering community on their efforts to affect the total cost of quality. A change in the total cost of quality or a shift from one quality cost category to the other would indicate to upper management, as well as to PA management, that PA had some effect on reducing variation and increasing the level of quality and reliability.
2. The relationship between the activities provided by PA (Q and R consultation, tool implementation, and analysis) and the reduction in variation and number of failures found in the plant or in the field helps direct the entire organization toward continuous improvement.
3. The goal of shifting the cost of quality from failure costs to preventive costs initiates the incentive of the PA organization to concentrate on Q and R activities during concept and early product design and process development. These activities would involve PAP, JIT training, and Q and R tool implementation and consultation.

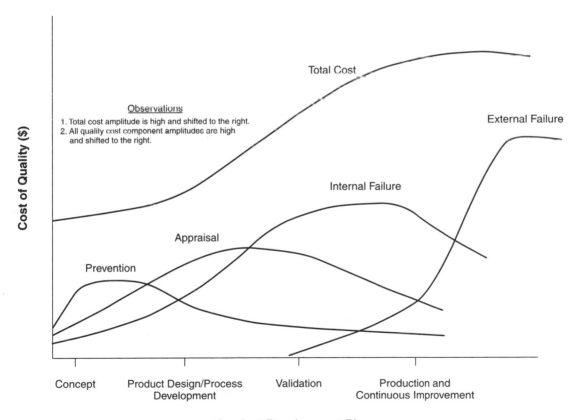

Figure 9.8b Quality Costs and the Product Development Process (Actual Condition)

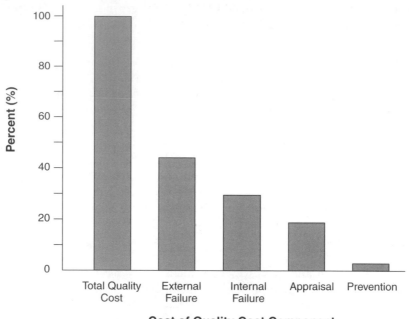

Cost of Quality Cost Component

Cost Component		Expense	Percent of Total Cost
Prevention		$1,000	2.6
Q & R Tools	$500		
Training	$500		
Appraisal		$8,000	21.1
Receiving Inspection	$3,000		
Validation Testing	$5,000		
Internal Failure		$12,000	31.6
Rework	$4,000		
Scrap	$8,000		
External Failure		$17,000	44.7
Warranty	$17,000		
Total Cost of Quality		$38,000	100%

Figure 9.9   Cost of Quality Management Example

4. The cost of quality and life-cycle analysis provides a predictive methodology for guiding the magnitude of product design and process improvement.
5. Cost of quality management promotes teamwork and communication between engineering management and other departments such as supplier quality, purchasing, etc.
6. Cost of quality management enables all of management to know the size of the quality and reliability opportunity and where to assert the most effort.

NOTES

1. Bieda, John, "A Product-Cycle Cost Analysis Process and Its Application to the Automotive Environment," *1992 Proceedings—Annual Reliability and Maintainability Symposium,* 1992, pp. 422–425.

10

Product Assurance Managerial Practices and Behavior

"Drive out fear, so that everyone may work effectively for the company."

> DEMING'S EIGHT POINT
> FRANK PRICE,
> *Right Every Time,*
> ASQ QUALITY PRESS, 1990, P. 134

"Remove the barriers that rob people in management and in engineering of their right to pride of workmanship. This means, *inter alia,* abolition of the annual or merit rating and of management by objective."

> DEMING'S TWELFTH POINT
> FRANK PRICE,
> *Right Every Time,*
> ASQ QUALITY PRESS, 1990, P. 134

Chapter Introduction

The primary purpose of this chapter is to describe useful product assurance managerial practices and behavior. These practices provide management with effective guidelines for managing their organization. Providing a friendly work environment and a team atmosphere helps employees strive for significant accomplishment in the organization. The guidelines covered in this section explain the techniques used to foster a constructive and satisfied product assurance organization.

Product assurance managerial practices and behaviors are important considerations to the effective management of a product assurance organization. These practices and behaviors should be carefully integrated into the main philosophy and task of a practical product assurance organization. As mentioned before, the tasks of a practical product assurance program involve implementation of the PAP, strategic use of Q and R tools during the PDP, just-in-time training, and cost of quality management. Specific managerial practices and behaviors are utilized to assure that the basic principles of product assurance management are utilized effectively to achieve the goal of improving or "building-in" quality and reliability into product designs and processes. Highlights of some of these managerial behaviors are shown in Figure 10.1.

A complete list and description of these practices and behaviors, which help focus the attention of product assurance into accomplishing the objectives of the main organization, can be summarized as follows:

1. *Require management support and the ability to offer advice on key managerial decisions regarding product design/process issues.* The PA organization must be viewed by the rest of the organization as a key activity that has significant leverage in product quality/reliability decisions. PA must have organizational visibility if people are to realize the systematic approach to "building-in" quality and reliability. A PA organization that is not accepted or utilized during product development, regardless of corporate programs they may establish, will not add value to the product and may be misinterpreted as just a cost burden.

2. *Emphasize to the organization through practice and presentation the value of strategically incorporating the various quality and reliability tools and processes to help design or improve product designs and manufacturing processes throughout the product development program.* Product assurance management should reinforce the application of the particular tool and its output by explaining in simple terms to those team members how the information resulting from the tool has added value to the product. It is important that PA management makes senior engineering management aware of this value and how it affects a product design or process risk.

3. *Encourage cooperation and teamwork with engineering and the supplier community on implementation of Q and R tools.* Product assurance should provide extra support to those engineering groups who need more confidence in understanding the tools which will aid them in product design and process development. Product assurance personnel must be ready to assist engineering and help provide explanation on the usage and interpretation of the results from each Q and R tool. The product assurance management should help promote group effort between various engineering groups in the problem-solving process through the use of these tools. The rest of engineering must be allowed to feel comfortable with these tools.

4. *Continuously demonstrate to upper management and engineering the relationship between the type of tools and techniques used to assess quality and reliability and the risks present in a particular design or process.* A demonstration of the correlation between the results of a tool and the potential product risk helps convince upper management and engineering how such a tool provides significant engineering value during a particular period in the product development cycle. This information involving risk and its relationship to the tool used to demonstrate it should be translated to the FMEA or FTA.

5. *Facilitate the transformation of design/process improvements made from the use of specific Q and R tools to the FMEA or FTA.* Product assurance management should emphasize to engineering the importance of using the FMEA as the central document for tracking improvements that have affected the risks in the product design and process. Using this tool as the central document helps to relate actual risks to engineering activity.

Prevention versus Reaction

- Promote preventive versus reactive product development practices.

- Stress the integrated use of product assurance tools and other engineering analysis techniques.

- Recognize and reward the individuals and the team for using various Q and R tools to impact the cost of quality and significantly improve quality and reliability.

- Enforce and praise the disciplined activity for reduction of product risk through the FMEA. Keep the FMEA as the central document of engineering activity performed to evaluate risk.

- Remove fear from the organization by making the organization understand and respect the PA function.

- Help the engineering community manage a *simple PAP* while focusing on risk evaluation and preventive engineering development.

Figure 10.1   Highlights of Preferred Product Assurance Managerial Practices and Behavior

6. *Offer the opportunity to chair design reviews and guide the process to addressing risks and formulating controls as recorded in the FMEA.* This opportunity supports the practical use of the FMEA or FTA in being the principal tool for communication in product design or process development. Product assurance management could help structure the design review to better address key questions regarding product quality and reliability.
7. *Facilitate the communication between design and process engineering through participation in the DFMA process early in the design/process development phase.* Product assurance management must stress the need to evaluate the product early in the PDP in order to eliminate risks in the manufacturability of the product.

8. *Praise and reward fellow PA engineers or other engineering personnel on the efforts and recorded success of preventing defects and reducing the cost of quality through specific uses of Q and R techniques.* Product assurance management must frequently advertise the success stories of PA preventive activities in order to promote belief, confidence, and motivation. A record of these success stories and relationships between the PA activity and the cost of quality should be made known to upper engineering management.

9. *Consult engineering on the program management of the PAP, but avoid micromanaging the activity.* Allow the engineering staff to use the PAP as a guideline with some mandatory requirement on specific product assurance activity. Product assurance management should continue to reinforce those activities in a PAP which significantly impact the quality and reliability assessment during product development. Product assurance should not overburden engineering with extensive planning as opposed to the appropriate use of specific Q and R tools, which produce the most benefit in the designated development period.

10. *Do not elicit PA involvement too late in a particular phase in the PDP.* Product assurance management should be active as early as possible in considering improvement through evaluation of the design or process. Product assurance should strive for product design and process variation reduction early in the product development process when the quality costs are smaller.

11. *Appraise PA employee performance on the use of the Q and R tool and the improvement of the product design or process.* Product assurance management must carefully examine the relationship of the execution of the PA activity through Q and R utilization to the improvement of the design or process. It is important to avoid comparison to other employees and to support team accomplishment toward improvement. In addition, PA management should not allow the main organization to misrepresent the role of PA or directly compare them to engineering groups. Upper management and product assurance management should frequently recognize the PA engineer and the PA organization for the team effort in using the specific Q and R tools to help improve product design/process quality and reliability, and lower the costs of quality.

12. *Continuously advocate and practice the use of Q and R tools on preventive activities of product development versus reactive activities.* Prevention activity through PA support offers greater rewards than reactive activity. Preventive activity results in lower costs of quality and increased customer satisfaction.

13. *Address questions and issues regarding product design and process development keeping in mind the guidelines as expressed in ISO 9000 and QS 9000 requirements.* Product assurance management should specifically focus on the questions as detailed in the Assessor's Guide of the QS 9000 document. These questions are addressed through the actions carried out by the product assurance organization. It is necessary to help the engineering community understand the correlation of the PA activities to the elements as described in the ISO/QS manuals.

14. *Maintain constancy of purpose through PA action and implementation of support activities such as PAP development, Q and R consultation and use of Q and R tools, training, and cost of quality management.*

15. *Stress lean production versus mass production philosophy to the organization through concentration of Q and R tool implementation during the design and process development phase.* Focus on communication, teamwork, and the reduction of variation through continuous improvement. Help product design and process engineering better understand and incorporate design/process improvements through the use of Q and R tools.

16. *Encourage the establishment of accelerated test investigation and development for product design/process testing.* Product assurance management should promote activity in the organization to obtain correlation of field failure and the type of accelerated test stress

environment. This task in correlating field failures to accelerated test stress environment would provide a better indication of product performance and verification in a shorter time frame.

17. *Promote training in engineering simulation with statistical support for all new product designs or processes.* Product assurance should be working closely with designers to help in the statistical interpretation of design simulations in order to avoid costly redesigns after prototyping due to dimensional incompatibility, material or electrical overstress, etc.

18. *Show no preference in resolving quality and reliability issues with any engineering function or affiliated organization such as purchasing, manufacturing, supplier development, etc.* Product assurance management should communicate their efforts to all of those engineering groups or affiliated organizations that will help them in their job functions.

19. *Avoid performance appraisal systems that instill fear and inaction within the PA organization or between PA and the engineering community.* Award the employee who works with the team, tries hard, and offers assistance wherever possible. Upper management should reward product assurance's effort to help prevent problems and not reward primarily on reaction.

20. *Assure that the PA organization is a recognized entity of the organization and is fully supported by the top management of the organization.* This means that management is willing to incorporate the decisions of PA into consideration of product design and process issues. In addition, the PA organization should be recognized by the talent of its people and the capacity in which they can serve the rest of the organization.

The previously mentioned PA managerial practices and behavior are important guidelines to consider when trying to satisfy primary objectives of the PA organization as well as the main goals established for the entire organization. The order in which to use these guidelines is variable, depending on the specific type of organization (manufacturing only, design and manufacturing, design only, etc.). Most guidelines should be considered at the beginning of a product program and implemented across all phases of a product development program since each guideline is not particularly attributable to one phase or another.

11

Future Direction of Product Assurance

"The industry is in continuous development, and so are the tempers of consumers. Both demand more and better quality."

SPOKESMAN FOR THE EGYPTIAN COTTON
EXPORTING COMPANIES, QUOTED IN
THE *New York Times*, JANUARY 15, 1971.

Chapter Introduction

The purpose of this chapter is to indicate future product assurance opportunities in the manufacturing and service-related industries. Each future development demonstrates to the reader the many challenges facing the quality or reliability professional in either manufacturing or service related-industries. Provided in this chapter are some examples of product assurance activities and issues which face the product assurance professional.

A. Growth in Design and Manufacturing

Product assurance is an integral engineering discipline to any organization and will have significant impact on the design and manufacture of a product. Major advances in product assurance which impact design and manufacturing include the direct integration of Q and R statistical tools with simulation and accelerated test methods. Both of these advances can be used early in the product development cycle to provide the greatest benefit in the development of a risk-free product design and processes. Other significant future events of product assurance include the integral relationship of PA management to upper management. The role of product assurance and the practical elements of its discipline should be clearly understood and implemented from top management on down. This integration of product assurance personnel at the upper management level of the organization will help facilitate the use of PA throughout all phases of product development and at all levels and departments of the organization. Product assurance will not just be a support group, but a normal function of a company's operations.

A specific description of the future developments in product assurance management for product design and manufacturing are detailed as follows:

1. *Focus on the integration of mechanical and electrical engineering simulation and statistical tool applications in quality and reliability engineering.* The combined approach of simulating structural, thermal, fluid dynamic, etc., stresses on a particular design using finite element analysis (FEA) and statistically evaluating the variation or time to failure helps to transfer product design improvement directly into the product. This approach may be specifically accomplished by simulating a stress distribution, evaluating the response per some specified design or process tolerance, and computing quality and reliability engineering statistics to determine design capability.

   Electrical simulation may be accomplished by stressing the power, voltage, or current through a circuit and translating this to a stress ratio of actual versus design. Failure rate models can be then used to incorporate the stress data and other environmental factors (thermal and vibration input information obtained through other simulations) to produce a reliability estimate. Directly studying the effects of mechanical and electrical stresses through simulation and statistical analysis allows for a better understanding of risk early in product development.

2. *Linking quality and reliability tools into a systematic product design/process development and improvement approach would help to identify risk and prevent the occurrence of high risk before verification testing.* This method utilizes reliability growth management and DOE processes in an iterative fashion to facilitate meeting desired reliability objectives and reducing high risks as identified in the FMEA (see Figures 11.1 and 11.2).[1]

3. *Combine the simulation of manufacturing process operations with statistical interpretation of process performance or capability at each station to help expedite design and quality analysis during process development.* Just as it was for the combined product design simulation and statistical analysis, manufacturing process development can be performed and evaluated more thoroughly before hard tools are produced and product is built through the production intent assembly line system.

4. *Encourage the product designer to work more directly with the product assurance engineer and incorporate full utilization of simulation techniques and requirements.* This kind of relationship would facilitate the development of reliable prototypes sooner and before hard tools are used to build the product on the actual assembly line.

5. *Incorporate virtual reality techniques and statistical analysis to improve understanding of product risk while actually viewing the physical responses from a three-dimensional perspective.*

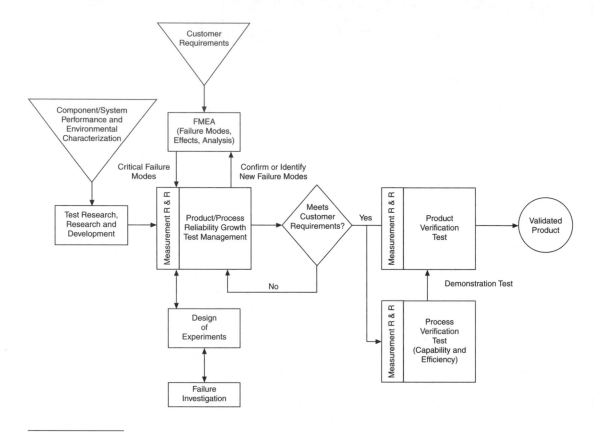

Source: John Bieda, "A Practical Product/Process Reliability Development and Improvement Approach," SAE Technical Series #932881, 1993, p. 54. Adapted by permission.

Figure 11.1   Product/Process Reliability Development and Improvement Approach

6. *Use accelerated testing techniques for development and verification testing.* The accelerated test approach would include an understanding of the environmental stress type, its magnitude, duty cycle, and the failure mode as compared to similar product field experience. This correlation of test failures and field failures is key to the development of the accelerated test profile. Future use of accelerated testing will involve a significant understanding of test stress to actual failure before test execution (see Figures 11.3a and 11.3b).

7. *Establish comprehensive databases containing such information as warranty, field failure history (outside of warranty), failure mode and mechanism (FMEA), internal defect failures, design guidelines, part duty cycle and environmental profiles, etc.* Future product assurance management will involve continuous use and application of data from these sources in order to develop representative tests to field exposure.

8. *Periodically review manufacturing maintenance records in order to evaluate level of equipment and tool maintainability.* Future applications in product assurance management would involve consultation with manufacturing on the repair records for new and used pieces of equipment. The capability of computerized, ongoing tool and equipment record collection and analysis to manage preventive maintenance is a significant activity in the reduction of process variation and internal failures.

9. *Advocate product design and process simulation over actual testing.* Product assurance management would be a key discipline in managing the balance between these efforts in product design and process development.

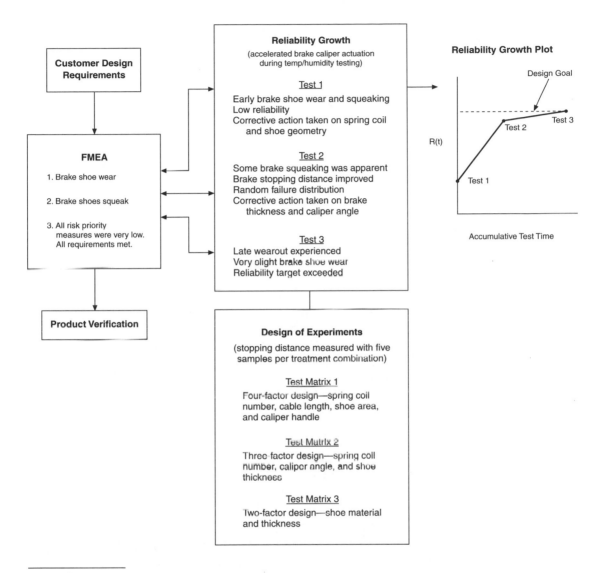

Source: John Bieda, "A Practical Product/Process Reliability Development and Improvement Approach," SAE Technical Series #932881, 1993, p. 65. Adapted by permission.

Figure 11.2 Reliability Development and Improvement Approach to the Design of a Bicycle Brake System—Example

10. *Emphasize the building of an Expert System to contain a body of knowledge regarding the part's performance history, packaging advantages, design/process requirements, material requirements, manufacturability, etc.* This Expert System would also have the potential to help recommend the preferred component or subcomponents needed to satisfy specific design and process requirements for the system.

11. *Focus on design for manufacturability and assembly right from the onset of product design and process development.*

12. *Encourage and support the use of product assurance concepts and tools throughout the organization, from senior management on down to the people working on the assembly line.* This transformation of product assurance principles will involve training at all levels as to how specific tools work and how the outputs from these tools translate to concise operator description sheets.

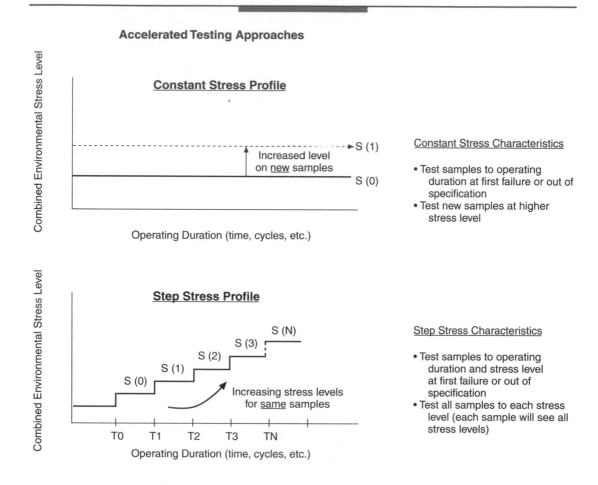

The main intent of accelerated life testing is to achieve an understanding of the inherent component design reliability and environmental stress applications responsible for particular failure mode mechanisms without incurring extensive laboratory test time. There are two basic methodologies to implement in executing and evaluating accelerated testing. They involve either the use of a constant or a step stress application. The attempt of either method is to achieve the following benefits: (1) verify product design reliability, (2) reduce test time by testing the samples to an elevated stress environment, (3) accelerate normal field failure mechanisms while not introducing new mechanisms, (4) compare and extrapolate test results at an elevated stress environment to normal use failure distributions, and (5) decrease cost by testing for a shorter time duration.

Figure 11.3a A Comparison of Accelerated Testing Approaches—Constant Stress versus Step Stress Techniques

A strategy for implementing and evaluating the results from either the constant or step stress approach is summarized below:

Constant Stress Method

1. Choose the environmental stress factor(s) types such as voltage, current, power, mechanical force, vibration amplitude/frequency, temperature, humidity, etc., and factor levels (low, medium, high), which may affect the functionality of the sample. It is advisable to use a designed experiment approach (full or fractional DOE techniques) to determine the significant stress types and levels.

2. Incorporate the environmental stress factor characteristics into a test profile and test each sample design to first failure at this constant environmental stress level.

3. Plot the failure distribution (cumulative percent failures versus cumulative test time—cdf) per failure mode and identify the three or two parameter Weibull characteristics from the failure distributions. A plot of the environmental stress and the product life characteristic can be constructed to illustrate the acceleration effect (see profile below).

4. Obtain warranty failure information from the field and separate these failure modes into distinct failure mode and failure mode mechanism categories.

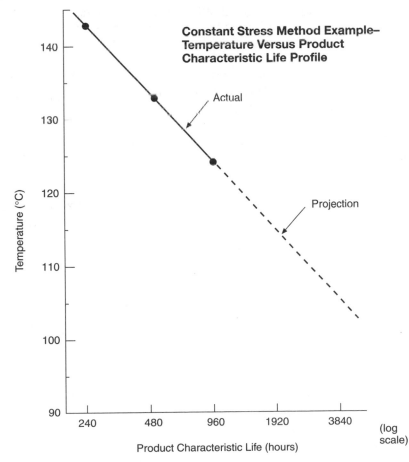

Figure 11.3b Constant and Step Stress Methodologies

5. Perform Weibull analysis for each field failure mode or mechanism and obtain the Weibull distribution characteristics such as slope (shape parameter), characteristic life (scale parameter), and minimum life (location parameter).

6. Compare the field and laboratory test failure mode Weibull characteristics. If the Weibull slopes are equal, formulate the ratio of the characteristic lifes for the laboratory and field failure modes. From this information, determine the field to laboratory test parameter time to failure relationship. If the Weibull slopes are unequal, calculate the laboratory-to-field ratio and use this conversion ratio to determine the reliability or cdf for both distributions. Next, take 50 percent of the unreliability and read off the corresponding conversion multiplier. Calculate the time equivalent to the conversion ratio.

7. Repeat steps 1 through 6 for other significant environmental stress factor combinations.

Step Stress Method:

1. Choose the most significant environmental factor(s) based on design of experiment investigation.

2. Test samples through the first level of the predetermined environmental stress factor combination. Record the first instance of failure for each sample at this level.

3. Use Weibull analysis or other reliability demonstration techniques to evaluate the probability of success. Refer to section B, Part 2 of Chapter 6 to understand some of the other reliability analysis methods.

4. Adjust the environmental stress factor combination to the next level and record the first instance of failure. *These samples should be the same samples that underwent testing at the previous stress level.*

5. Determine the posterior (conditional) distribution using Bayesian statistics. From this conditional probability formulation a probability of success can be computed.

6. Continue to test at high stress levels until all samples have exhibited first failure. Calculate the final conditional probability of success using Bayesian statistics.

Figure 11.3b *Continued.*

13. *Focus on systems engineering and analysis through simulation and testing.* Future developments in product assurance management will involve evaluating the risk to the system and its interfaces to components through more comprehensive system models and information about preestablished components. Techniques may involve frequent use of fault tree analyses to quickly identify fault paths and responsible components. This systems engineering approach would specifically follow a series of steps (see Figure 11.4) and include: (1) recognition of system boundaries, (2) identification of system requirements,

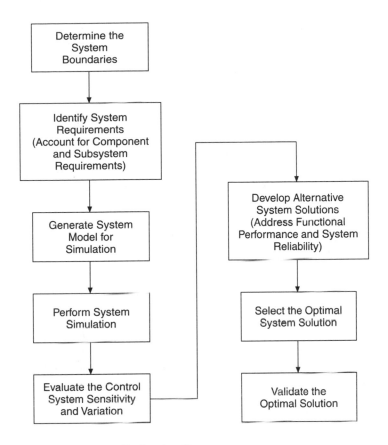

Figure 11.4   System Engineering Evaluation Process

(3) development of a system model, (4) simulation, (5) sensitivity analysis, (6) evaluation of alternative solutions, (7) determination of optimal solution, and (8) validation of the optimal solution.

14. *Consistently use cost of quality management, from concept development through the production phase of the PDP.* Future applications will involve a computerized link of the cost impact from the efforts of PA and other departments of the organization to help manage where effort should be provided to reduce the total cost of quality.

15. *Link the efforts of PA to the requirements set forth through QS 9000.* Regularly compare and contrast the PA efforts throughout the PDP to the basic elements described in the QS 9000. Benchmark the progress of the organization and its suppliers on the key questions posed in each element of the QS 9000.

16. *Standardize PAP usage and application throughout the organization.* Future direction will involve common use of a practical format of the PAP on all components and systems, whether hardware or software.

B. Product Assurance Management in the Service Industry

Future developments in product assurance management of the service industry will parallel some of the activities previously described for design and manufacturing industries. Although, not explicitly emphasized in this book, the practical product assurance management approach is somewhat similar for service as it was for products. The use of a PAP, training, and cost of quality management is the same as it was for design and manufacturing industries. However, the implementation of specific Q and R tools will be different since the product under evaluation is a service as opposed to a physical product design. Such tools used in concept development (market analysis, QFD, test sample planning), product design and process development (simulation, design reviews, process control plan, etc.), validation (test result analysis), and production (Pareto charts, root cause analysis) would be quite similar for service analysis. The differences in the use of these tools would be in the specific treatment of the service procedure and parameters used to quantitatively and qualitatively assess the performance of the service to a specified customer requirement level. For example, the parametric distributions to be evaluated and the measurement and validation techniques would be different.

The product or service development process phase is where differences in the application of specific Q and R tools is most apparent. Specific techniques which would require slightly different approaches would be in simulation, development testing, and measurement system capability. The service procedure can be described, timed, and measured, and the quality of product evaluated by the customer. Given the information, a simulation of the desired output service parameter can be studied for a multitude of input conditions. The significant inputs which affect the output can be reviewed and reexamined for improvement if needed. In regard to testing, trial runs of the service practice can be observed for different customer conditions and reviewed according to the criteria and measurement parameters. The analysis would either involve attribute or variable type data and would concentrate on variances of the desired service practice to a standard. Finally, the measurement system would be very sensitive to the parameter chosen in a service industry operation, especially if it is operator sensitive such as visual inspection. However, as in some product hardware cases, attribute gauge R and R techniques could be used to evaluate the measurement system.

C. Growth in the Service Industry

The need for product assurance management is critical for many types of services, particularly in a society where service businesses are increasing at a faster rate over the number of design and manufacturing companies. Service industries that are experiencing the need for a practical product assurance program would include the medical profession (hospitals, outpatient clinics), banking firms, retail outlets, product service establishments (auto repair, appliance repair), etc. Quality and reliability are concepts which can be applied and evaluated in every commodity or profession where there is a need for improvement. Several quality and reliability objectives that are considered in the service industry are: (1) identification of the service requirement to be monitored, (2) the risk of the particular service requirement, and (3) how the service requirement can be measureed and validated in a repeatable manner. An example of a risk analysis approach, as utilized in design and manufacturing, can be applied in the medical profession (see Figure 11.5).

The growth foreseen in the service industry would be centered around activity during the early concept development phase. The task involved in this phase would be the ability to recognize all the customer requirements desired and how these requirements translate into measurable parameters which may be evaluated and benchmarked as a function of the change in the operating environment or customer population. Since the requirement definition activity in a service development program is the most fundamental entity prior to any development, it will demand the first and most effort before the main elements of a product assurance program can be a factor in bringing forth significant improvement in the organization.

A list of service product assurance opportunities for the present and near future include the following:

1. Understanding the relationship of QS-9000 elemental objectives as they relate to the particular service procedure(s).

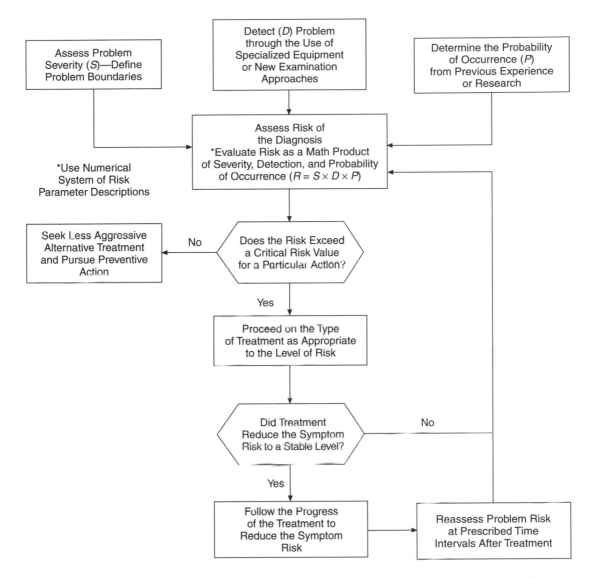

Figure 11.5 Product Assurance Risk Reduction Technique—A Medical Treatment Example

2. Identification of the product assurance activities and tools to be used for the service PDP—a service PAP.
3. Determination of the service product development program dates and milestones for a typical service is the first and mandatory step.
4. A change in the mentality of upper management in understanding the need for a product assurance program of a service. It is necessary to reinforce the idea that there is no difference in the pursuit of quality and reliability for a service as compared to a product.
5. Identification of the quality cost elements and the analysis procedure for the management of the cost of quality for the particular service.
6. Determination of the relationship between the tools and processes used to evaluate the service operation (service flow diagram, risk analysis—FMEA or FTA, service control plan,

preventive maintenance plan, and operator instruction sheet). Establish how the requirement parameter is measured and controlled throughout the service operating period.

7. Focus on measuring and evaluating variances in a defined requirement such as billing amounts, materials, time of service, number of repeat occurrences or problems, etc.

8. Incorporation of lean production philosophy with emphasis on teamwork, communication, continuous improvement, and inventory control.

NOTES

1. Bieda, John, "Practical Product/Process Reliability Development and Improvement Approach," SAE Technical Paper Series, Worldwide Passenger Car Conference and Exposition, Oct. 1993, pp. 53–66.

12

A Summary of Practical Product Assurance Management

"Put everybody in the company to work in teams to accomplish the transformation."

DEMING'S FOURTEENTH POINT
FRANK PRICE,
Right Every Time,
ASQ QUALITY PRESS, 1990, P. 154

Chapter Introduction

This last chapter provides a summary of the principal benefits of practical product assurance management. These benefits are a selection of the most important advantages found in the four functional elements covered in this book. All key benefits discussed in this chapter are significant qualities which add value to the management of a product assurance organization and the company in which this organization serves. The chapter is concluded with some thoughts on practical product assurance management.

A. Benefits

The benefits of practical product assurance management do significantly impact the quality and reliability for a particular product design and process during the product development process. The institution of a product assurance organization which facilitates teamwork and communication between engineering groups through the incorporation of timely simultaneous engineering techniques is a valuable entity to an organization striving to achieve continuous improvement and a lower cost of quality. Allowing the product assurance activity to affect product development can be simply summarized by: (1) instituting PAP to guide the product quality and reliability development in the PDP, (2) training personnel on a just-in-time basis on fundamentals of the Q and R tools to be used, (3) implementing the use of quality and reliability tools to detect defects, obtain root cause, and prevent failures, and (4) assessing cost of quality, helps toward the goal of "building-in" high quality and reliability into the product.

The following is a list of the benefits of practical product assurance management in an organization:

1. Guides product development thinking toward strategic use of PA tools to help "build-in" quality and reliability. These tools detect potential failures or unnatural variation, facilitate selection of alternative designs and processes based on predicted performance in the field, and promote preventive measures for assuring design performance and process control.
2. Promotes an understanding of the relationship of the PA assessment to the total cost of quality by correlating the way tools were used by PA during each phase of the PDP to the magnitude of the failure, appraisal, and prevention cost components of the total cost of quality.
3. Encourages communication and teamwork between engineering disciplines (design, manufacturing, materials, tooling, etc.) and other organizational groups such as purchasing, value engineering, product planning, etc. PA bridges the gap of relating product design/process performance and cost to reliability and quality by identifying areas of quality and reliability opportunity.
4. Helps direct the use of training and execution of Q and R tools to assessing risk in a timely manner through the PDP. Product assurance management becomes the key element to shifting the cost of quality from a failure-oriented activity to a preventive activity.
5. Links the key elements of Deming's fourteen points on quality and those requirements set forth through the QS-9000 document.
6. Helps increase management's awareness of the magnitude of risk and the opportunity for improvement throughout product development. PA facilitates upper management's ability to incorporate probabilistic data in its decisions.
7. Offers a cost/benefit alternative to traditional reactionary behavior activity of a quality engineering department that is solely responsible to product design/process problems after they occur. Practical PA management offers proactive activity to the detection and prevention of product design and process problems.
8. Provides insight into product design and process evaluation for in-house product engineering personnel and supplier personnel during a specific point in the product development process.

B. Concluding Remarks

The need for practical product assurance management is important from the perspective of moving toward prevention versus reaction and the obtainment of more reliable and higher quality designs and processes. Executing particular PA tools and processes in a very simple and logical manner during specific phases of a product development cycle is a primary goal of practical PA management. A practical product assurance management program essentially involves the strategic use of the four elements explained in this text: (1) PAP, (2) just-in-time training, (3) application of various Q and R tools, and (4) cost of quality assessment. Managing these four key elements in the PDP so as to help detect, evaluate, and correct potential problems early is the essence of practical product assurance. Product assurance should not be complex in what it is to an organization or in how the tools are implemented. Instead, PA should be a simple *process*, which is clear to the organization as to what, why, who, when, where, and how it significantly impacts product design and process quality and reliability. The focus of PA is not just setting requirements or solving problems once the product is in production, but rather, a carefully orchestrated use of combinations of tools applied in a simple manner to assess risk and promote product improvement prior to verification.

This book was written to explain the use of four principal elements and how they are utilized to "build-in" quality and reliability into the product design and process. The general philosophy that these four elements emphasize is the ability to shift the improvement from later in the PDP where changes and corrections are expensive and more risky to customer satisfaction, to early in the PDP where all engineering disciplines can carefully evaluate the opportunities for improvement as the design and process evolve from customer requirements. The PAP organizes the strategic incorporation of Q and R tools by addressing the what, where, who, and when these tools should be used during the PDP. Just-in-time training helps the team better recognize the primary application of the various tools just before they are applied on actual products. Proper execution of several of the various Q and R techniques in a systematic manner, by using the results of one tool to support the other tool, helps to better indicate where potential problems are and what type of redesign or improvement is necessary to simplify or correct the product. Finally, the last PA principle of cost of quality management directs the attention of management as to where the source of waste is, or the product risk in financial terms, and how the total cost of quality measures up to the bottom line—profits and customer satisfaction. When all of these elements are applied in syncronicity through the PDP, the net effect is a shift in product improvement upstream and a reduction in the total cost of quality.

The implementation of product assurance management requires a complete commitment of the organization, starting with top management. Its success in the organization depends on the willingness of management to constantly support and reward for the preventive actions conducted by the team as opposed to reactive actions.

Glossary

accelerated testing   A technique of increasing the stress on a component or system to accelerate normal use failure mechanisms without introducing new mechanisms. Accelerated testing would help to reduce test time, decrease cost, and enable reliability verification. Two types of accelerated testing are considered: constant stress and step stress. Multiple or combined stress environments are used in accelerated testing but must be treated carefully due to the concern of sorting out multiple failure mechanisms or introducing new mechanisms.

acceptance/rejection criteria   Performance specifications used to determine the success or failure of a particular characteristic.

acceptance sampling plan   A description in tabular or graphical format of the sample size needed to determine acceptance or rejection for a specified lot of material and AQL (Average Quality Level). This acceptance plan may also define sample sizes based on the severity classification for the particular product.

accuracy   A specific parameter for the assessment of a measurement's system calibration. Accuracy is the difference between the observed average of measurements and some master value which is traceable to an official standard.

analysis of variance (ANOVA)   An estimation technique which apportions the total variation of a set of data into individual component parts associated with various sources of variation for the primary purpose of testing hypotheses on the parameters of some model.

appraisal costs   A quality cost component associated with verifying design or process quality. These costs are associated with such activities as inspection, testing, acceptance sampling, auditing, etc. This quality cost component is one of three major components of the total cost of quality.

attribute data   Qualitative data that can be counted for recording and evaluation. The occurrence of a defect based on some acceptance/rejection criteria would be an example of this kind of data.

average and range control chart (X-bar and R chart)   A variable data control chart method for monitoring inspected data in terms of its location (process average) and spread (range—part to part variation). The process average chart or X-bar is specifically defined as the average of values in small subgroups over an operating duration. The range chart is defined as the range of values within each subgroup (high value minus low value) over an operating duration. These charts are used to evaluate the behavior of data relative to upper and lower control limits. A data point outside of these limits or a special cause of variation based on the trend of the data denotes a possible out-of-control condition, which should warrant further investigation. Both of these variable data control charts comprise a method out of several others used in the application of statistical process control.

average and range (long) method   A mathematical approach to determining the total measurement system repeatability and reproducibility (total measurement system variation), and the contributors to this total variation—equipment and appraiser variation. Total measurement system repeatability and reproducibility (R and R) as a percent of part tolerance is usually applied for new processes without prior history while total R and R as a percent of process variation is applied for established processes that have included process improvement.

225

average quality level (AQL)   The maximum percentage of unacceptable samples in a lot that, for the purposes of acceptance sampling, can be considered satisfactory as the process average. This AQL value serves as an index for acceptance sampling plans.

Bayesian method   A process of estimating the probability of the causes that may have produced an observed event. In other words, the Bayesian technique enables a determination of the probability for an event B, or cause, that may have brought about the event A. This Bayesian process may involve an approximation of the probability of an event in question given limited data based on knowledge of prior probabilistic information.

benchmarking   A methodology of measuring your company's processes by those of the best competitors to help establish a standard for performance. Benchmarking is used to help learn about other related processes, adapt these methods, and take appropriate action to meet and exceed the standard in order to become the industry leader. Benchmarking consists of four basic steps: (1) preparation, (2) fact finding (measurement and analysis), (3) action plan development, and (4) maturity and recalibration.

binomial distribution   A probability distribution based on the *Bernoulli process,* which involves an experiment that can result in one of two mutually exclusive outcomes, correct or incorrect, and is performed through sampling *with replacement* (i.e., sampling from an infinite population). This process is based on the condition of mutually exclusive outcomes, a probability of success that remains constant, and an independence of one outcome to the other.

calibration   The process of determining the capability of a measurement system through comparison of established measurement standards. Its purpose is to determine the

magnitude of measurement resolution through accuracy and linearity. Both attribute and variable type data measurement systems involve a calibration determination.

chi-square distribution   An estimation of the variance for a normally distributed population using the sampling distribution of $(n - 1)s^2/\sigma^2$. The larger the number of samples drawn (n), the closer the graph of the empirical sampling distribution is to a chi-squared (χ^2) distribution. Unlike the normal or t distributions the chi-square distribution is asymmetric. There is a different distribution for each possible value of degrees of freedom, $n - 1$.

chi-square test   A type of goodness-of-fit test used for sample sizes greater than 25 for comparing the expected frequency distribution with the corresponding observed frequency distribution. The decision determined in this test of hypothesis involves comparing a test statistic, based on the expected and observed values of the frequency distribution, to some critical value per a certain level of significance.

classical design of experiments   A predetermined series of tests in which multiple factors are evaluated simultaneously to determine their individual and combined influence on a output variable or response. The total number of tests or treatment combinations is a mathematical function of the number of factors and the levels of each factor. Unlike the Taguchi design of experiments, the classical DOE includes all treatment combinations.

concept approval   The engineering milestone period when a decision is made as to the product design intent.

confidence   The likelihood that a probabilistic statement made is true. For example, "90 percent reliability at 90 percent confidence" means that there is 90 percent certainty that the probabilistic claim (i.e., reliability) is indeed a true statement.

cost breakdown matrix   Direct repair costs, failure rates, and production volume information necessary to evaluate the unreliability and subsequently the number of estimated incidents per an operating duration.

cost impact summary matrix An estimate of the total cost of various design alternatives in order to allow for management decision making at various phases of the product development process. The total cost amount represents the sum of the piece cost and life-cycle cost per a particular product operating period.

cost of quality management The process of managing the individual cost of quality components (prevention, appraisal, and failure costs) and life-cycle cost for a particular product during the various phases of the product development process. It involves the efforts of various groups in an organization so as to enable production and service at the most economical level that allows for full customer satisfaction.

customer satisfaction The ability to properly accommodate all customer expectations in a timely manner, with low risk of failure, and at a reasonable expense.

Deming's fourteen points A listing of key elements that an organization should consider to improve the quality of the system in which they are currently operating. Deming had formulated these fourteen points in the 1970s to provide direction and focus toward a quality system. These points cover such areas as continuous improvement, teamwork, communication, training, management behavior, etc.

descriptive statistics Basic statistical indicators of central tendency and variation of sample data distributions. Some of these statistical indicators are the mean, variance, standard deviation, etc.

design failure mode and effect analysis (DFMEA) A formalized method for quantifying the risk associated with identified potential design failure modes and their failure mechanisms and effects. It is used to help identify design controls necessary to prevent failures from occurring. This DFMEA contains a risk priority number for each failure mechanism based on a mathematical product of a probability of occurrence ranking, severity ranking, and a detection ranking.

design for manufacture and assembly (DFMA) A technique that facilitates the quantification of risks involved in various assembly alternatives. It involves combining the activities of product design as well as the manufacturing processes including assembly operations into a multifunctional simultaneous engineering activity. The analysis of a DFMA focuses on such guidelines as a minimum number of parts, ease of handling, symmetrical parts, etc.

design of experiments (DOE) A statistical methodology for efficiently planning and analyzing a test to determine the most significant factors contributing to the greatest influence of a desired output variable. It essentially involves planning the test, setting it up, running it, analyzing the data, and verifying the results.

design reliability growth test management A process for identifying failure modes, incorporating design changes, and monitoring reliability progress on an ongoing basis during the design/process development phases of a product development process. The methodology involves testing samples and monitoring the reliability or MTBF as a function of some accumulative test duration. A comparison of the test-fix-retest process to some target growth line is accomplished using one of a variety of reliability growth models such as the Duane growth model (for exponentially distributed failures).

design reviews Structured meetings with representatives from applicable areas that assess and review results of an activity directed toward product design/process development. These meetings may involve review of design/process requirements, test results, design/process studies, and functional objectives. Design reviews are held periodically throughout the product development process and before formal decision points (prior to engineering milestone periods) in a program.

design varification tests (DV) Tests performed during the validation phase of the product development process to help verify that the final released component, subsystem, or system meets the design intent requirements. The particular test types, test sequences, sample sizes, acceptance/rejection criteria, etc., are detailed in product performance standards or highlighted in product test sample reports.

direct repair cost matrix A matrix containing costs associated with removal and replacement, repair, shipping, processing, and development. These costs are used to facilitate life-cycle cost analyses.

Duane growth model A model that is closely related to the Weibull reliability growth model and is defined as the accumulative failure rate being equivalent to a constant (depending on equipment complexity, design margins, etc.), the accumulative test time t, and the slope of the growth curve (see article by Peter Mead, "The Role of Testing and Growth Techniques in Enhancing Reliability," 1978).

duty cycle The number of product operating cycles per some specified operating duration. Statistical descriptions of qualitative or quantitative data representing expected customer usage level (i.e., 50th, 90th and 95th percentile) and environmental profiles over the life expectancy for a particular product application. An example of a duty cycle may be the number of electrical relay cycles in 500 hours of continuous operation.

engineering development tests (ED) A series of systematic tests performed during the product design/process development phases and prior to the validation phase of the product development process to help establish design parameter thresholds and targets, verify conformance to requirements, and determine reliability.

engineering milestone A scheduled, engineering decision event prior to the approval of a particular set of development activities.

error/mistake proofing A method that involves the ability to design out possible errors and mistakes from a design or process. Error proofing is a method used to identify potential product design/process errors and either design them out or eliminate the possibility that the error would ever produce a defect in the first place. Mistake-proofing is a method applied to identify errors that may occur and prevent them from becoming a defect that could reach the customer. The mistake-proofing activity may involve changing a process operation that could help the operator become more aware of the error potential and prevent the future occurrence of a mistake.

F-test A statistical test of variances (square of the standard deviation), which helps determine if a significant change in the shape of the measurement distribution occurred. The F-test is based on a distribution known as the F distribution and is determined when the sample variances are calculated from samples that have been randomly and independently drawn from normally distributed populations with equal variances. The appropriate F distribution is determined from the degrees of freedom associated with the numerator and denominator of F. A decision as to whether the two population variances are not equal (reject null hypothesis that variances are equal) depends on whether the computed test statistic exceeds the critical value.

failure costs A quality cost component associated with internal and external defects that are experienced either during development, production, or in the field. Examples of internal failure costs are the result of scrap, rework, repair, retest, etc. Examples of external failure costs are the result of warranty replacement, products rejected and returned, etc. This quality cost component is one of three major components of the total cost of quality.

fault tree analysis (FTA) A structured top-down approach used to evaluate the risk of a particular failure event. This method involves the use of Boolean algebra and symbols to diagram the fault path of failure modes leading to the top-level failure event.

finite element analysis (FEA) An engineering technique for mathematically describing through a model a complex structure as a collection of elements which are interconnected with a definite number of nodes. When this model is simulated to known loads (mechanical, thermal, etc.), the functional response of the structure may be determined through the solution of a set number of equations which account for element interaction.

first time capability (FTC) The initial probability of success assessment of those parts successively processed through an operation or total manufacturing process. Other manufacturing process indices may be used for variable data type process parameters.

fishbone diagram A tree diagram of possible causes leading to some particular effect. The possible causes identified in the tree diagram become factors to be studied further for level of contribution to the effect under investigation.

Fussel Vesely (FV) importance measure A relative index that defines the fractional contribution of a fault tree event to the risk. It is specifically defined as the ratio of the difference of the current risk and the decreased risk level with the feature/function optimized to the present risk.

functional block diagram An illustration of the functional relationship of various components, subsystems, or systems for the purpose of engineering evaluation during the product development process. This diagram should illustrate the type of functional characteristics (mechanical, electrical, etc.) of each component or subsystem and the nature of the interconnection of one component or subsystem to the other.

gauge repeatability and reproducibility An analytical method used to evaluate a measurement system's magnitude of total variation as a function of tolerance or process control limits. This total variation is comprised of repeatability and reproducibility. Repeatability is a measure of a measurement system's ability (gauge, etc.) to repeat a given measurement and is evaluated by repeatably measuring the same parts (equipment variation—EV). Reproducibility is a measure of the ability of people to reproduce a series of measurements using the same gauge and is evaluated by two or more operators independently repeating the inspection of a set of parts (appraiser variation—AV).

goodness-of-fit Tests of hypotheses to evaluate any distribution assumption using quantitative data. These tests specifically involve a comparison of a set of observed frequencies with their expected (or theoretical) frequencies. Some of the popular goodness-of-fit tests are the chi-square test (25 or more samples), Kolmogorov-Smirnov test (25 or fewer samples), Hollander-Proschan (compares the theoretical reliability function of the Kaplan-Meier estimate of the reliability function), etc.

hardware Pieces of mechanical or electrical equipment that provide fit, form, and function. This equipment may provide direct function or may even provide mechanical or electrical function to other pieces of equipment.

hypergeometric distribution   A probability distribution used when sampling is performed *without replacement* from a finite population and it is desired to determine the probability of some particular number of acceptable or unacceptable outcomes. This probability distribution is utilized when the binomial distribution is not applicable.

inherent reliability   The designed-in reliability or a product's intrinsic reliability under field and/or stated conditions for a specified design life.

instantaneous failure rate   The incremental number of failures per incremental operating duration. The failure rate is defined as the probability density function divided by the reliability.

instantaneous mean time between failure (MTBF)   The incremental average time between failure. MTBF is defined as the reciprocal of the failure rate.

just-in-time   The ability to manufacture or service (training) only what is needed, when it is needed, and in the amount needed. This concept is very appropriate to inventory systems where it is necessary to minimize the amount of inventory on hand so as to reduce total cost (through the cost of holding inventory) and convert those materials into finished goods.

key characteristics   Parameters applicable to a component, material, manufacturing, or assembly operation which are critical to part fit, form, function, appearance, and having particular quality/reliability significance.

Kolmogorov-Smirnov test   A type of goodness-of-fit test used for sample sizes less than 25 for comparing the expected frequency distribution with the corresponding observed frequency distribution. The decision determined in this test of hypothesis involves comparing the maximum value for the difference between the expected and observed values of the frequency distribution and some critical value per a certain level of significance.

lean production   A philosophy which is firmly based on teamwork, communication, continuous improvement, and inventory control. The efforts of a lean production organization promote improvement in quality, inventory control, utilization of manufacturing space, reduction in the number of engineering changes and the cost of change, etc.

life-cycle cost analysis (LCC)   A process of evaluating the short- and long-term product effectiveness through a combination of reliability engineering and cost accounting techniques. The LCC methodology involves combining unreliability data with apportioned direct repair costs to produce a cost of repair estimation for some particular product operating duration.

life expectancy   The specific time or duration for the useful functional operation of a system, component, or material.

maintainability   The probability of successfully performing and completing a maintenance action to restore a failed system to operational readiness in a specified interval of downtime. Downtime is the total time in which the particular system is out of functional service and involves failure detection time, logistic time, repair time, etc.

manufacturing floor plan   An illustrative layout of the entire manufacturing facility. It should include identification of equipment and machinery, testing space, office space, shipping and receiving areas, nonconforming material areas, etc.

market research   The process of identifying which kinds of products, types of features, and functions satisfy customers' expectations. This process of understanding customer expectations may involve customer surveys, questionnaires, interviews, etc.

mass production   The traditional philosophy which focuses on high productivity only at the expense of teamwork, communication, quality, and inventory control. Mass production represents the style of organizational behavior brought on by the advent of Henry Ford's manufacturing and assembly line operation in the beginning of the twentieth century.

maximum likelihood   A best-fit curve fitting approach used in Weibull analysis to generate the Weibull slope, which favors those data points that occurred later in the experiment. This method is an iterative procedure of solving the maximum likelihood equations.

mean time to repair   The average duration necessary to restore equipment and tools to operational readiness.

method of least squares   A statistical estimator of the maximum likelihood through determining the minimization of the sum of squares for the deviations between the independent and dependent variables. The simple linear regression model can be used to generate the equation (see Bernard Ostle, *Statistics in Research,* Second Edition, 1963).

normal distribution   The most important and commonly used distribution in statistics. It is symmetrical about its mean μ. The mean, median, and mode of this distribution are all equal. Other characteristics of this distribution include a total area under the curve above the *x*-axis equal to one (50 percent of the area is to the right of a perpendicular line drawn at the mean, and 50 percent is to the left). The area is equated to: (1) +/– one standard deviation from the mean is equal to 68 percent of the total area, (2) +/– two standard deviations from the mean is equal to 95 percent of the total area, and (3) +/– three standard deviations from the mean is equal to 99.7 percent of the total area.

on-target   A statistical process performance or process capability index for assessing the magnitude of the sample data mean coinciding with the nominal specification value. This measure of being "on-target" is specifically defined as the difference between either P_p or C_p and P_{pk} or C_{pk}, respectively. The index is not used for one-sided tolerances, that is, specifications with only a minimum or only a maximum. When evaluating the "on-target" index, a value that is smaller is better and zero is best (sample mean exactly coincides with the nominal specification value).

operator instructions   Documented forms posted at each process operation to help describe and visually present assembly steps and key processing concerns. These instructions will also contain information regarding inspection requirements, process controls, gauge and tool procedures, SPC instructions, acceptance/rejection criteria, etc. The operator instructions are the result of the effort initiated from the process flowchart, PFMEA, process control, and preventive maintenance plans.

packaging   The type of package material and package design utilized in the storing and shipping of parts. The package design provides protection and containment of the part. It affects ease of handling by manual or mechanical methods. Packaging may consist of returnable or nonreturnable items.

Pareto chart   A graphical reporting technique for displaying attribute (counting) data where the greatest frequency is displayed to the left and decreasing frequencies to the right. The chart represents a comparison of the number of times particular events have occurred.

planning and concept development phase A set of engineering activities or tasks (assignments) performed to accomplish conception and approval of a particular design. This phase involves such activities as market research, product requirement definition, test plan development, etc.

prevention costs A quality cost component associated with preventing defects in the product design or manufacturing process. These costs are associated with such activities as quality and reliability engineering training, designed-in quality and reliability, a quality system program, quality planning (product assurance planning—PAP), etc. This quality cost component is one of three major components of the total cost of quality.

preventive maintenance (PM) A series of actions conducted according to a periodic and planned schedule to retain a product in a specified working condition through checking and reconditioning. These actions are precautionary activities performed to reduce the tendency for failure or an unacceptable condition.

process A combination of people, equipment, methods, materials, and environment that facilitates the development of a product.

process capability indices Statistical measures of the potential and capability of a manufacturing process to perform within specified limits. C_p, C_{pk}, and $C_p - C_{pk}$ are used when volume production data is available for an extended period. Process potential, C_p, is defined as the ratio of the difference in the upper and lower tolerance limits to the process variation ($6\sigma_R/d_2$). The larger the value for the process potential indice, the greater the ability of the distribution to fit inside the tolerance limits. Process capability, C_{pk}, is defined as the ratio of the difference between the sample distribution mean and the closest tolerance limit to the process variation ($3\sigma_R/d_2$). The larger the value for the process potential indice, the greater the ability of the sample distribution mean to locate near the nominal specification value. The measure of a process being "on-target", $C_p - C_{pk}$, is defined as the difference between the process potential and capability. This measure of being "on-target" is a relative indicator of whether the sample distribution is centered about the nominal specification value. The smaller the difference between C_p and C_{pk}, the better the sample distribution mean is centered about the nominal specification value.

process capability studies The activity of quantifying a process's inherent variation for *statistically stable processes* by using indices that help estimate the relationship of a sample variable data distribution to specified tolerance limits. Process potential and capability indices are used to indicate the magnitude of variation relative to these tolerance limits. The estimation of the standard deviation is defined as $\bar{R}/d_2$ and is based on the actual process behavior.

process control plan A detailed document used to help describe a system for controlling production parts and processes. It is written to address the important product design/process characteristics and engineering requirements of the product. This plan contains such elements as the process description, acceptance/rejection criteria, sampling size and frequency, measurement tools, control methods, and corrective action.

process failure mode and effects analysis (PFMEA) A formalized method of quantifying the risk associated with identified potential process failure modes, their mechanisms, effects, and the controls taken to prevent the probability of failure occurrence. The PFMEA rating system involves a risk priority number based on the mathematical product of the probability of occurrence ranking, the severity ranking, and the detection ranking. This risk priority number is used to prioritize the effort toward evaluating risk prevention.

process flow diagram An illustration indicating the steps in a manufacturing or assembly process from incoming material to packaging of the finished product. It includes process control stations, inventory, and material handling.

process performance indices   Statistical measures of the ability of a manufacturing process to perform within specified limits. P_p, P_{pk}, and $P_p - P_{pk}$ are used prior to production and help to provide a preliminary indication of process potential and capability. The process performance indice, P_p, is defined as the ratio of the difference in the upper and lower tolerance limits to the estimated process variation (6s). The larger the value for the process potential indice, the greater the ability of the distribution to fit inside the tolerance limits. The other process performance indice, P_{pk}, is defined as the ratio of the difference between the sample distribution mean and the closest tolerance limit to the estimated process variation (3s). The larger the value for the process potential indice, the greater the ability of the sample distribution mean to locate near the nominal specification value. The measure of a process being "on-target", $P_p - P_{pk}$, is defined as the difference between both performance indices. This measure of being "on-target" is a relative indicator of whether the sample distribution is centered about the nominal specification value. The smaller the difference between P_p and P_{pk}, the better the sample distribution mean is centered about the nominal specification value.

process performance studies   A *preliminary* evaluation of a processes' ability to produce a product within specified tolerance limits. The process performance indices specifically help to indicate the magnitude of variation relative to tolerance limits. These indices differ from process capability indices in the estimation of the sample standard deviation. Process capability indices are based on *statistically stable* processes, whereas process performance indices are not. The estimation of the sample standard deviation used in the process performance indices is defined as "s."

process reliability   The study of the probability of success for manufacturing process operations performed under various conditions for a designated operating duration.

The calculation of the process reliability would involve similar analytical approaches as for determining product design reliability. A determination of the characteristics of the data distribution for the data (attribute or variable type) collected for a particular process would initiate the calculation of the reliability. Total process reliability would involve the same calculation approach using reliability diagrams as for design reliability.

process reliability growth test management A method for identifying failure modes, incorporating process changes, and monitoring reliability progress on an ongoing basis during the process development phase of the product development process. This methodology involves assembling samples to a production representative process and monitoring the process performance or capability as a function of the product development milestone. A comparison of the test-fix-retest process to a process growth line or target process performance level allows for relative measure of process improvement over real time.

process review   A systematic review of a particular manufacturing process (in-house manufacturing and assembly, supplier, etc.) conducted by various members of the product team. This review involves examination of process documentation and an actual evaluation of the manufacturing process operation.

process verification testing (PV)   Tests performed in the validation phase of the product development process to validate design fit, form, and functionality of initial production-line-made parts manufactured from production intent tools and assembly processes.

product assurance   An organization of quality and reliability engineering professionals who incorporate various Q and R tools and processes during the product development process to improve the product or service and help *prevent potential failure*. This organization works along with the product design/manufacturing organization to fully develop a product. It is a kind of organization that helps make important decisions during the early phases of product development and facilitates the teamwork and communication of various disciplines involved throughout the product development process.

product assurance planning (PAP)   A method of identifying product development tasks necessary to facilitate the accomplishment of key engineering events during a product development process.

product design development phase   A set of engineering activities or tasks (assignments) performed to develop and evaluate the conceived design from the planning and concept development phase. This phase involves such activities as simulation, engineering development tests, risk analysis, etc.

product development process (PDP)   A series/parallel arrangement of product development phases involved in the design and manufacture of a particular product or service. The phases in the development process may involve concept planning, design development, process development, validation, production, and continuous improvement.

product process development phase   A set of engineering activities or tasks (assignments) performed to develop and evaluate the manufacturing process originally conceived from the planning and concept development phase. This phase involves such activities as simulation, risk analysis, process control plan development, measurement systems evaluation, process performance studies, preventive maintenance plan development, etc.

product validation phase   A set of engineering activities or tasks (assignments) performed to verify the product design and manufacturing process as developed during the product design and process development phases. This phase involves the evaluation of results from design and process verification

testing as well as review of the production intent manufacturing process.

production and continuous improvement phase   A set of engineering activities or tasks (assignments) performed to monitor and improve the production intent manufacturing process as was validated in the product validation phase. This phase utilizes Pareto charts, statistical process control methods, root cause analysis, etc. to help identify and correct product design/process problems.

proportion nonconforming control chart (p-chart)   An attribute control chart method for monitoring inspected data in terms of the proportion of non-conforming items in a group. The p-chart helps to monitor the behavior of the data relative to specified control limits. A data point outside of these limits or a special cause of variation based on the trend of the data denotes a possible out-of-control condition, which should warrant further investigation. This attribute data control chart methodology is one method out of several for evaluating attribute data in the application of statistical process control.

QS-9000 (quality system requirements)   A manual developed by Chrysler/Ford/General Motors and published by the Automotive Industry Action Group (AIAG) to describe key elements required for development of quality systems that provides for prevention, waste reduction, waste elimination, and continuous improvement. These key elements are also applicable to any design/manufacturing organization (other than the automotive industry) that is developing a quality system.

quality   The conformance to specific customer needs or requirements for the purpose of obtaining customer satisfaction.

quality and reliability tools   A collection of engineering and statistical techniques applied strategically to detect potential problems and facilitate the improvement of product design and manufacturing processes or service-related operations.

quality function deployment (QFD) An organized methodology to transform and then prioritize customer expectations and requirements into specific engineering design and manufacturing process requirements. This QFD approach provides a helpful analytical process for focusing the engineer to specific design/process requirements which can be developed into a product and validated.

range (short) method An abbreviated mathematical approach to determining the total measurement system repeatability and reproducibility. This technique, unlike the average and range method, does not include a calculation of the appraiser and operator variation contribution to the total repeatability and reproducibility. Instead, this method involves a calculation of the total repeatability and reproducibility. Total measurement system repeatability and reproducibility (R and R) as a percent of part tolerance is usually applied for new processes without prior history, while total R and R as a percent of process variation is applied for established processes that have included process improvement.

rank regression A best-fit curve fitting approach used in Weibull analysis to generate the Weibull slope, which favors those data points that occurred early in the experiment. This method is based on simple linear regression model equations.

reliability The probability that a product will successfully perform as intended without failure, under specified operating and duty cycle conditions, for some specified duration (time, cycles, hours, etc.), at a given confidence level.

reliability "bathtub" curve A graphical representation of the relationship for a component, subsystem, or system's failure rate and operating duration. This pictorial representation shows a decreasing function or infant mortality (Weibull slope < 1), followed by a random failure behavior (Weibull slope = 1), and then concluded by an increasing function or wear-out behavior (Weibull slope > 1).

reliability block diagram An illustration of the individual component or subsystem reliabilities in some series/parallel arrangement for the purpose of total system reliability calculation.

reliability demonstration An execution and analysis of a test for measuring a product's conformance to specified design requirements, at a given confidence level, and relative to some acceptance/rejection criteria. The demonstration target is expressed as a reliability at a specified confidence level such as 90 percent reliability at a 90 percent confidence level (R90/C90). This R90/C90 would equate to either 22 samples successfully functioning through the required duration, or some sample amount of parts surviving past the required operating duration and resulting in a reliability (once an analysis using some statistical distribution—i.e., Weibull analysis) that meets or exceeds the target.

reliability prediction A statistical inference method of estimating reliability through the use of several reliability engineering methodologies from actual test data, failure rate models, etc. Reliability prediction can be made for systems, components, and materials for the purpose of addressing concept selection proposals, life-cycle cost estimates, etc.

reliability testing An activity of simulating field environments on samples to determine the probability of success relative to all functional requirements. Reliability testing may incorporate the use of accelerated testing methodologies in order to expedite the testing process.

risk priority number (RPN) A relative measure of the magnitude of design or process risk and a function of the probability of occurrence, severity, and detection of a particular failure mode. The risk priority number is the mathematical product of the probability of occurrence ranking, the severity ranking, and the detection ranking. The probability of occurrence ranking is a relative measure of the likelihood that a specific cause/mechanism will occur. The severity ranking is a relative measure of the seriousness of the effect of the potential failure mode to the next component, subsystem, or system. The detection ranking is a relative measure of the ability of the current design or process controls to identify a potential cause before the component, subsystem, or system is produced and sent to the customer.

root cause analysis   An organized approach of describing a problem, developing possible causes, evaluating possible causes, and confirming the main cause. This process utilizes various quality and reliability tools to help review failure mode information (Pareto charts, fishbone diagrams, FMEA), evaluate test or field results, or conduct designed experiments to identify contributors to a specific problem.

simultaneous engineering   The application of concurrent product design and process developmental activities to assist design and manufacture of a product. The use of cross-functional organizations is necessary in facilitating this process.

software   A set of written logical statements and instructions which control the function of equipment, machines, computers, etc. These logic statements may be written using different computer languages that specifically describe logical operations.

specifications   Specific engineering requirements for assessing the acceptance/rejection criteria for product design/process fit, form, or function. These specifications may be described through variable or attribute data means. The use of variable data limits may be composed of upper and lower limits, or either a minimum or maximum limit. The use of a description for

identifying acceptable product design/process quality (appearance, form, etc.) may be used to describe attribute type data specification.

stable processes   Conditions of statistical control through minimum variation exhibited from common cause as opposed to special causes of variation. For process performance studies the determination of a stable process may not be as well defined as for actual production capability studies since the sample size and production build period are not extensive.

statistical control   A condition when a process exhibits no special causes of variation but only common causes of variation. This determination of statistical control may be achieved through the use of process control charts to help indicate the relationship of data points to specified control limits. The trend in the data with respect to these control limits is an important part of determining the magnitude of process variation or statistical control.

statistical process control (SPC)   A formalized method of utilizing process control charts to monitor specific process parameter variation within specified control limits. These statistical process control studies are performed on variable and attribute type data to assess the degree of statistical control.

stress/strength interference analysis   A method to evaluate the relationship of the stress and strength distribution and the probability of failure (interference). The magnitude of interference between the stress and strength distribution defines the actual probability of failure. The strategy for achieving a reliable design is to increase the distance between the stress and strength distributions, which would produce sufficient safety margin from potential failure.

system reliability   The mathematical determination of total reliability from some series/parallel combination of component or subsystem reliabilities. The arrangement of the complex series/parallel component reliabilities is based on a functional block diagram for the system under investigation.

t-test   A statistical test performed between two means to determine if a significant change exists in the mean value of the measurement distributions. The t-test is based on the student's *t* distribution and is determined for a number of samples of size *n* from a normally distributed population using the mean and standard deviation. Some characteristics of the *t* distribution are that it has a mean of 0, is symmetrical about its mean, has a variance greater than 1, and approaches the normal distribution as *n* increases. The t-test operates by comparing a test statistic to a critical value at some particular level of significance in order to determine whether the sample means are indeed different (reject the null hypothesis that the means are equal). *It should be emphasized that a type of t-test used for comparison in this book will treat the data as independent samples and will evaluate sensitivity between the two means based on the following criteria: (1) the effects of the various stresses are unknown prior to test and may affect the uniform treatment assumption of the paired t-test, (2) the t-test of independent samples has greater sensitivity than the paired t-test for small sample sizes, (3) the risk of error is small; an incorrect decision will merely result in a higher level of scrutiny to determine the practical significance.*

Taguchi design of experiments (DOE)   A specific application of classical or factorial DOE in which the effect caused by one of the many experimental factors in a designed experiment can be separated from the effects of the remaining factors. This method of experimental design is described through the use of orthogonal arrays which are arithmetic means balanced, not mixed or confounded. The DOE matrix which is formed by these orthogonal arrays allows a smaller number of tested to be evaluated, but sufficient coverage of the total number possible factor/level combinations.

test data analysis   A series of statistical techniques applied to determine various

descriptive statistics, test of hypotheses, probabilities, etc., for sample data distributions. This analysis may involve attribute (go–no-go) and variable (measured values) type data. These kinds of analyses may be performed to make decisions with statistical confidence.

test effectiveness   A measure of overall software test coverage from a theoretical total number of input combinations. This measure is specifically defined as a ratio of the actual number of software input combinations tested from the DOE matrix and the theoretical total number of input combinations. Software test effectiveness can be related to some measure of confidence.

tests of hypotheses   Statistical tests of validity for an assertion made about a population which are performed by analysis of sample data. These tests of hypotheses involve determining whether the sample data represent a distribution for a particular population. These tests are specifically performed by comparing a test statistic to a critical value and either accepting or rejecting a claim to a distribution type. In evaluating a hypothesis, two errors can be made: type I error (reject the hypothesis when it is true) or type II error (accept the hypothesis when it is false).

tool and equipment studies   The evaluation and determination of the life expectancy and repair frequency necessary for use in a planned preventive maintenance study. These studies would include review and evaluation of the time to repair data for various perishable tools and equipment using traditional statistical analysis techniques.

treatment combinations   How the factors and levels of each factor are configured to establish a test.

type I error (level of significance)   The probability of rejecting a hypothesis when it is true. This error is typically denoted by α and is often referred to as the producer's risk.

type II error   The probability of accepting a hypothesis when it is false. This error is typically denoted by β and is often referred to as the consumer's risk.

variable data Measurable data for analysis purposes. Examples of this type of data may be mechanical fit dimensions, electrical characteristics, mass, volume, etc.

variation simulation analysis (VSA) A technique used to statistically evaluate the variation of an output response variable based on a function of specific input variables developed through a mathematical algorithm. It uses the distributions of input data characteristics and a model to define the relationships for critical fit, form, or function parameter.

Weibull distribution technique A commonly used distribution in reliability engineering to help identify characteristics of various data distributions. The characteristics that the Weibull distribution technique helps to identify are the shape (Weibull slope), scale (characteristic life), and location (minimum life) parameter. Each of these parameters is used to determine the probability density function (pdf) of a given data distribution. From the pdf the failure rate and reliability distribution equations can be formulated and applied to describe the behavior of product or service performance.

zone of operation Specific operating ranges that software code can be exercised with combinations of inputs using boundary analysis to identify whether the software can handle the various zone inputs.

Index

A

Analysis of risk
  failure mode and effect analysis, 66–72
    application
      target point for, 67
      timing of, 67
    benefits of, 68–72
    implementation
      individual responsible for, 68
      technique, 68
    selection of, 66–67
  fault tree analysis, 72–76
    application
      target point for, 73
      timing of, 73
    benefits of, 74–75
    implementation
      individual responsible for, 73
      technique, 73–74
    selection of, 72–73
Application of product assurance, within
      process, 8–9
Approval process
  production part, 21
Assembly, manufacturability, design
      for, 76–79
  application
    target point for, 76
    timing of, 76
  benefits of, 78–79
  implementation
    individual responsible for, 77
    technique, 77–78
  selection of, 76
Assembly plant performance characteristics,
      compared, 14
Audit, of quality, internal, 21
Automotive product development
      performance, comparison, 15

B

Barriers, removing, as product assurance
      technique, 19
Barriers between departments, break
      down of, 19
Bayesian quantification, of past experience, in
      concept development, 54
Benchmarking, 11, 38–44
  action plans for results of, 43
  analysis, 42–43
    matrix example, 43
  application
    target point for, 39
    timing of, 38–39
  benefits of, 43–44
  design
    monitoring action taken, 43
    strengths, weaknesses, 42
  example, 40–42
  implementation technique, 39–43
  information, collection of, 42–43
  process
    monitoring action taken, 43
    strengths, weaknesses, 42
  selection of, 38
  steps, 39
Binomial distribution, sample size
      determination, 52
Block diagrams, 64–66
  application
    target point for, 64
    timing of, 64
  benefits of, 65–66
  flow chart, 65
  functional, 64–65
  implementation
    individual responsible for, 64
    technique, 64
  selection of, 64

C

Capability, manufacturing, product assurance
     and, 21
Capability analysis, process performance and,
     148–58
  application
    rationale for, 148–49
    target point of, 149
    timing of, 149
  benefits of, 157–58
  implementation, 149–57
  technique, 149
Carlyle, Thomas, 32
Company organization, relationship of
     product assurance activity to, 9
Complexity level, product assurance
     involvement, 8
Concept development, 32–62
  analysis, 44–45
  Bayesian quantification, past experience, 54
  benchmarking, 38–44
    action plans for results of, 43
    analysis, 42–43
     matrix example, 43
    application
     target point for, 39
     timing of, 38–39
    benefits of, 43–44
    design
     monitoring action taken, 43
     strengths, weaknesses, 42
    example, 40–42
    implementation technique, 39–43
    information, collection of, 42–43
    process
     monitoring action taken, 43
     strengths, weaknesses, 42
    selection of, 38
    steps, 39
    tool, selection of, 38
  benefits of, 54–55
  Carlyle, Thomas, 32
  product assurance planning and, 26–27
  quality function deployment, 33–38
    application
     target point for, 33
     timing of, 33
    benefits of, 38
    chart creation, 34–37

    example, 36, 37
    house of quality matrix, 33
    implementation
     individual responsible for, 33
     technique, 33–38
    phases of, 34
    selection of, 33
    technique, 35
  reliability demonstration, tests used for,
     49–54
  requirement definition, 33–38
  sample planning, and analysis, 44–45
  sample size requirements, reliability
     demonstration, 54
  test plan development, 44–45, 55
    benefits of, 62
    hardware model, 44
  test profile example, 46
  test sample planning, example, 50–51
  verification test sequence, example, 47
Constancy of purpose, as product assurance
     technique, 19
Continuous improvement, 21, 31, 173–88
  Pareto chart analysis, 182–84
    application
     target point for, 183
     timing of, 183
    benefits of, 184
    example, 183
    implementation
     individual responsible for, 183
     technique, 183–84
    selection of, 182
  root cause analysis, 184–88
    application
     target point for, 184
     timing of, 184
    benefits of, 188
    example, 186–87
    implementation, individual responsible
     for, 184–85
    process, 185
    selection of, 184
    technique, implementation, 185–88
  statistical process control, 174–82
    application
     rationale for, 174
     target point for, 174
     timing of, 174
    benefits of, 182

control charts, table of constants, formulas for, 179
implementation
individual responsible for, 174
technique, 174–82
Contract review design control, as product assurance technique, 21
Control
contract review design, as product assurance technique, 21
of customer supplied product, 21
Corrective action, as product assurance technique, 21
Cost of quality management, 2, 5, 20, 195–204
elements, components of, 197
example, 204
life-cycle cost analysis, 198–202
example
cost breakdown matrix, 200
cost impact summary matrix, 200
direct repair cost matrix, 199
flowchart, 199
key elements provided, 201
Main, Jeremy, 195
tasks associated with, 203–4
Customer supplied product, control of, 21

D

Data control, 21
Delivery, as product assurance technique, 21
Deming, W. Edwards
on change, 1
fourteen points of, 18–20
on teams, 23
Demonstration test analysis, 134–38
application
rationale for, 134
target point of, 134
timing of, 134
benefits of, 138
chi-square method, 134
implementation, 134–38
technique, 134
Weibull analysis method, 134
Departments, barriers between, break down of, 19
Dependence, on mass inspection, 19
Design
development of, 63–112

for manufacturability, assembly, 76–79
application
target point for, 76
timing of, 76
benefits of, 78–79
implementation
individual responsible for, 77
technique, 77–78
selection of, 76
review. *See* Design review
validation of, 113–72. *See also* Validation
Design of experiments, 11, 158–63
application
rationale for, 158
target point of, 158
timing of, 158
benefits of, 162–63
implementation, 158–62
technique, 158
Design review, 11, 109–12
application
target point for, 110
timing of, 110
benefits of, 112
implementation
individual responsible for, 110–11
technique, 111
selection of, 109–10
DFMA. *See* Design, for manufacturability and assembly
Diagrams, block, 64–66
application
target point for, 64
timing of, 64
benefits of, 65–66
flow chart, 65
functional, 64–65
implementation
individual responsible for, 64
technique, 64
selection of, 64
Distribution analysis, Weibull's, 138–42
application
rationale for, 138
target point of, 138
timing of, 138
benefits of, 142
implementation, 139–42
technique, 139
Document control, 21

Documentation
  as product assurance technique, 21
  progression, quality system, 4
DOE. *See* Design of experiments

E

Education, as product assurance technique, 19
Effectiveness, of product assurance, during
      product development, 7
Eliminate numerical goals, as product
      assurance technique, 19
Engineering, in product assurance planning, 24
Equipment, test, 21
Experiments, design of, 11, 158–63
  application
    rationale for, 158
    target point of, 158
    timing of, 158
  benefits of, 162–63
  implementation, 158–62
    technique, 158

F

Failure mode and effect analysis, 10, 66–72
  application
    target point for, 67
    timing of, 67
  benefits of, 68–72
  implementation
    individual responsible for, 68
    technique, 68
  selection of, 66–67
Fault tree analysis, 10, 72–76
  application
    target point for, 73
    timing of, 73
  benefits of, 74–75
  implementation
    individual responsible for, 73
    technique, 73–74
  selection of, 72–73
FEA. *See* Finite element analysis, 25
Fear, removing, as product assurance
      technique, 19
Finite element analysis, 25
  in product assurance planning, 25
FMEA. *See* Failure mode and effect analysis
Fourteen points, of Deming, 18–20
FTA. *See* Fault tree analysis
Future developments, product assurance, 210–20

G

Growth, in design, manufacturing, 211–17

H

Handling, as product assurance technique, 21
Hardware
  assurance
    defined, 3
    relationship to product development
      process, 5
  reliability of, vs. software reliability, 4
History, of product assurance, 13–22
House of quality matrix, 35

I

Identification, of product, 21
Improvement, continuous, 21, 173–88
  Pareto chart analysis, 182–84
    application
      target point for, 183
      timing of, 183
    benefits of, 184
    example, 183
    implementation
      individual responsible for, 183
      technique, 183–84
    selection of, 182
  product assurance planning and, 31
  root cause analysis, 184–88
    application
      target point for, 184
      timing of, 184
    benefits of, 188
    example, 186–87
    implementation, individual responsible
      for, 184–85
    process, 185
    selection of, 184
    technique, implementation, 185–88
  statistical process control, 174–82
    application
      rationale for, 174
      target point for, 174
      timing of, 174
    benefits of, 182
    control charts, table of constants,
      formulas for, 179
    implementation
      individual responsible for, 174
      technique, 174–82

Initial process performance, capability studies, 11
Inspection
    mass, ceasing dependence on, 19
    as product assurance technique, 21
Internal quality audits, 21

J

JIT training. *See* Just-in-time training
Jones, Daniel T., 13
Just-in-time training, 2, 5, 17, 20

K

Kolmogorov-Smirnov test for normality, 119–23

L

Lean production philosophy, 14–18
    characteristics of, 16
    product assurance management principles,
        relationship, 17
Life-cycle cost analysis, 198–202
    example
        cost breakdown matrix, 200
        cost impact summary matrix, 200
        direct repair cost matrix, 199
    flowchart, 199
    key elements, 201

M

Main, Jeremy, 195
Maintenance, preventive, 98–103
    application
        target point for, 98
        timing of, 98
    benefits of, 101–3
    implementation
        individual responsible for, 98–99
        technique, 99–101
    life studies, 100–102
    scheduled plan, 101
    selection of, 98
Management
    practices of, 205–9
    of product assurance, overview, 1–12
    responsibility of, 21
Manufacturability, assembly, design for, 76–79
    application
        target point for, 76
        timing of, 76

benefits of, 78–79
    implementation
        individual responsible for, 77
        technique, 77–78
    selection of, 76
Manufacturing capabilities, product assurance
        and, 21
Mass inspection, ceasing dependence on, 19
Mass production philosophy, 14–18
    characteristics of, 15
Measurement system evaluation, 90–98
    application
        target point for, 91
        timing of, 91
    benefits of, 97–98
    calibration, 91–92
    gauge repeatability, reproducibility, variable
        data measurement systems, 92–96
    implementation
        individual responsible for, 91
        technique, 91–97
    selection of, 90
Measuring, as product assurance technique,
        21
Medical services, product assurance
        management in, 217–18

N

Need for product assurance management, 5–6
New philosophy, adoption of, 19
Nonconforming product, control of, 21

O

Operation, philosophy of, 3
Organizational structure, 9
    relationship of product assurance
        activity to, 9

P

Packaging, as product assurance
        technique, 21
PAP. *See* Product assurance plan
Parallel verification test sequence, series, 47
Pareto chart analysis, 182–84
    application
        target point for, 183
        timing of, 183
    benefits of, 184
    example, 183

implementation
    individual responsible for, 183
      technique, 183–84
  selection of, 182
Part performance characteristics, compared, 14
Philosophy, of operation, 3
  adoption of, 19
Preventive maintenance, 98–103
  application
    target point for, 98
    timing of, 98
  benefits of, 101–3
  equipment, life studies, 100–102
  implementation
    individual responsible for, 98–99
    technique, 99–101
  plans, 11
  as product assurance technique, 21
  scheduled plan, 101
  selection of, 98
  tool, life studies, 100–102
Price, award of business on, ceasing, 19
Process control plan, 85–90
  application
    target point for, 86
    timing of, 86
  benefits of, 90
  implementation
    individual responsible for, 87
    technique, 87–90
  selection of, 85
Process development, 27–30, 63–112
  block diagrams, 64–66
    application
      target point for, 64
      timing of, 64
    benefits of, 65–66
    flow chart, 65
    functional, 64–65
    implementation
      individual responsible for, 64
      technique, 64
    selection of, 64
  design review, 11, 109–12
    application
      target point for, 110
      timing of, 110
    benefits of, 112
    implementation
      individual responsible for, 110–11
      technique, 111

    selection of, 109–10
  manufacturability, assembly, design for, 76–79
    application
      target point for, 76
      timing of, 76
    benefits of, 78–79
    implementation
      individual responsible for, 77
      technique, 77–78
    selection of, 76
  measurement system evaluation, 90–98
    application
      target point for, 91
      timing of, 91
    benefits of, 97–98
    calibration, 91–92
    gauge repeatability, reproducibility, variable data measurement systems, 92–96
    implementation
      individual responsible for, 91
      technique, 91–97
    selection of, 90
  preventive maintenance, 98–103
    application
      target point for, 98
      timing of, 98
    benefits of, 101–3
    equipment, life studies, 100–102
    implementation
      individual responsible for, 98–99
      technique, 99–101
    scheduled plan, 101
    selection of, 98
    tool, life studies, 100–102
  process control plan development, 85–90
    application
      target point for, 86
      timing of, 86
    benefits of, 90
    implementation
      individual responsible for, 87
      technique, 87–90
    selection of, 85
  product assurance involvement, 8
  reliability growth management, 103–9
    application
      target point for, 103
      timing of, 103
    benefits of, 109

implementation
 individual responsible for, 104
 technique, 104–9
 selection of, 103
risk analysis
 failure mode and effect analysis, 66–72
 application
  target point for, 67
  timing of, 67
 benefits of, 68–72
 implementation
  individual responsible for, 68
  technique, 68
 selection of, 66–67
 fault tree analysis, 72–76
 application
  target point for, 73
  timing of, 73
 benefits of, 74–75
 implementation
  individual responsible for, 73
  technique, 73–74
 selection of, 72–73
variation simulation analysis, 80
 application
  target point for, 80
  timing of, 80
 benefits of, 83–84
 implementation
  individual responsible for, 80
  technique, 81–86
 selection of, 80
Process performance and capability analysis,
 148–58
 application
  rationale for, 148–49
  target point of, 149
  timing of, 149
 benefits of, 157–58
 implementation, 149–57
  technique, 149
Process review, validation, 31, 113–72. *See also*
 Validation
 application
  target point for, 164
  timing of, 164
 benefits of, 171
 cause, 170–71
 corrective action procedure, 170–71
 description, 164
 error proofing, 169

failure mode and effect analysis, 164
flow diagram, 164
implementation
 individual responsible for, 164
 technique, 164–71
machine instructions, 170
material acceptance plan, 164–66
measurement systems plan, evaluation
 studies, 167
operator, instructions, 170
packaging, shipping, of parts, 170
parts handling plan, 170
preventive maintenance plan, 166
process performance, capability studies, 169
process verification tests, 171
product assurance plan, 169
selection of, 163
statistical process control, 166
tooling, equipment studies, 166
Product assurance planning, 2, 5, 18, 23–31
 concept development, 26–27
 continuous improvement, 31
 defined, 2–5
 Deming, W. Edwards, on teams, 23
 Deming's fourteen points, relationship, 19
 design development, 27
 finite element analysis, 25
 organization, steps in forming, 11–12
 process development, 27–30
 process verification, 31
 product assurance method, relationship, 26
 product development process, 24
  engineering milestones, 25
 production, 31
 simultaneous engineering, 24
 worksheet, 28–29
Product complexity level, product assurance
 involvement, 8
Production, continuous improvement, 173–88
Purchasing, as product assurance
 technique, 21
Purpose, constancy of, 19

Q

QFD. *See* Quality function deployment
QS-9000 requirements, relationship to
 product assurance, 21–22
Quality audits, internal, 21
Quality engineering evaluation,
 validation, 148–58

design of experiments, 158–63
    application
        rationale for, 158
        target point of, 158
        timing of, 158
    benefits of, 162–63
    implementation, 158–62
        technique, 158
    process performance and capability
        analysis, 148–58
    application
        rationale for, 148–49
        target point of, 149
        timing of, 149
    benefits of, 157–58
    implementation, 149–57
        technique, 149
Quality function deployment,
        benchmarking, 33–38
    application
        target point for, 33
        timing of, 33
    benefits of, 38
    chart creation, 34–37
    example, 36, 37
    house of quality matrix, 33
    implementation
        individual responsible for, 33
        technique, 33–38
    phases of, 34
    selection of, 33
    technique, 35
Quality tools, strategic implementation, 18–19

R

Rationale, product assurance management, 5–6
RCA. *See* Root cause analysis
Records, quality, control of, 21
Reliability demonstration
    sample size requirements, 54
    tests used for, 49–54
Reliability growth management, 103–9
    application
        target point for, 103
        timing of, 103
    benefits of, 109
    implementation
        individual responsible for, 104
        technique, 104–9
    selection of, 103

Reliability test data evaluation, validation
    block diagram analysis, 142–48
        application
            rationale for, 142–43
            target point of, 143
            timing of, 143
        benefits of, 148
        implementation, 143–48
            technique, 143
        types of, with mathematical
            representations, 144
    demonstration test analysis, 134–38
        application
            rationale for, 134
            target point of, 134
            timing of, 134
        benefits of, 138
        chi-square method, 134
        implementation, 134–38
            technique, 134
        Weibull analysis method, 134
    stress/strength interference, 131–38
        application
            rationale for, 131
            target point of, 132
            timing of, 131
        benefits of, 132–34
        implementation, 132
            technique, 132
    Weibull distribution analysis
            technique, 138–42
        application
            rationale for, 138
            target point of, 138
            timing of, 138
        benefits of, 142
        implementation, 139–42
            technique, 139
Reliability tools, strategic implementation,
        18–19
Removing barriers, as product assurance
        technique, 19
Responsibility, of management, 21
Retraining, as product assurance technique, 19.
        See also Training
Review of design, 11, 109–12
    application
        target point for, 110
        timing of, 110
    benefits of, 112

implementation
  individual responsible for, 110–11
  technique, 111
selection of, 109–10
Risk analysis, development of
  failure mode and effect analysis, 66–72
    application
      target point for, 67
      timing of, 67
    benefits of, 68–72
    implementation
      individual responsible for, 68
      technique, 68
    selection of, 66–67
  fault tree analysis, 72–76
    application
      target point for, 73
      timing of, 73
    benefits of, 74–75
    implementation
      individual responsible for, 73
      technique, 73–74
    selection of, 72–73
Roos, Daniel, 13
Root cause analysis
  production, continuous improvement,
    184–88
    application
      target point for, 184
      timing of, 184
    benefits of, 188
    example, 186–87
    implementation, individual responsible
      for, 184–85
    process, 185
    selection of, 184
    technique, implementation, 185–88

S

SA. *See* Simulation analysis
Sample, size requirements, reliability
    demonstration, 54
Sample planning, 44–45
Series parallel arrangement, verification test
    sequence, 47
Service industry
  growth in, future of, 218–20
  product assurance management in, 217–18
Servicing, as product assurance technique, 21

Simulation analysis
  application
    target point for, 80
    timing of, 80
  benefits of, 83–84
  implementation
    individual responsible for, 80
    technique, 81–86
  selection of, 80
Software
  assurance
    defined, 3
    relationship to product development
      process, 5
  reliability of, vs. hardware reliability, 4
SPC. *See* Statistical process control
Standards, of work, elimination of, 19
Statistical process control
  production, continuous improvement,
    174–82
    application
      rationale for, 174
      target point for, 174
      timing of, 174
    benefits of, 182
    control charts, table of constants,
      formulas for, 179
    implementation
      individual responsible for, 174
      technique, 174–82
Storage, as product assurance technique, 21
Stress/strength interference, 131–38
  application
    rationale for, 131
    target point of, 132
    timing of, 131
  benefits of, 132–34
  implementation, 132
  technique, 132
Structure, of product assurance, overview, 1–12
Supervision, modern methods of, 19

T

Test data analysis, 114–27
  attribute type data, 127–31
    application
      rationale for, 127
      target point of, 127–28
      timing of, 127

basic proportion technique, 128
binomial distribution technique, 128–30
  implementation, 128
    technique, 128
demonstration, 134–38
  application
    rationale for, 134
    target point of, 134
    timing of, 134
  benefits of, 138
  chi-square method, 134
  implementation, 134–38
    technique, 134
    Weibull analysis method, 134
variable type data, 114–27
  application
    rationale for, 114
    target point of, 114
    timing of, 114
  benefits of, 127
  implementation, 114–27
    technique, 114
Test development, 44–45, 55
  benefits of, 62
  hardware model, 44
Test equipment, and product assurance, 21
Test sample planning, 11
  example, 50–51
Test status, as product assurance technique, 21
Thomson, William, 113
Timing, of product assurance
    implementation, 6–7
Traceability, as product assurance
    technique, 21
Trainer, 192
  qualifications of, 193
Training, in product assurance, 189–94
  concepts, 191–92
  materials, types of, 191
  modern methods of, 192
  performance of, 193–94
  process, guidelines, 193–94
  reasons for training, 192
  target point, for training, 192
  timing of, 192
  tools, 191–92
  trainer. *See* Trainer
Transformation, as team effort, 19

V

Validation, design/process
  process review, 163–71
    application
      target point for, 164
      timing of, 164
    benefits of, 171
    cause, 170–71
    error proofing, 169
    failure modes and effects analysis, 164
    flow diagram, 164
    implementation
      individual responsible for, 164
      technique, 164–71
    material acceptance plan, 164–66
    measurement systems plan, evaluation
      studies, 167
    operator, instructions, 170
    packaging, shipping, of parts, 170
    parts handling plan, 170
    preventive maintenance plan, 166
    process control plant, 164
    process performance, capability
      studies, 169
    product assurance plan, 169
    selection of, 163
    statistical process control, 166
    tests, 171
    tooling, equipment studies, 166
  quality engineering evaluation, 148–58
    design of experiments, 158–63
    application
      rationale for, 158
      target point of, 158
      timing of, 158
    benefits of, 162–63
    implementation, 158–62
      technique, 158
    process performance and capability
      analysis, 148–58
    application
      rationale for, 148–49
      target point of, 149
      timing of, 149
    benefits of, 157–58
    implementation, 149–57
      technique, 149

reliability test data evaluation, 131–42
  block diagram analysis, system
      reliability, 142–48
  application
    rationale for, 142–43
    target point of, 143
    timing of, 143
  benefits of, 148
  implementation, 143–48
    technique, 143
  types of, with mathematical
      representations, 144
reliability demonstration test
      analysis, 134–38
  application
    rationale for, 134
    target point of, 134
    timing of, 134
  benefits of, 138
  chi-square method, 134
  implementation, 134–38
    technique, 134
  Weibull analysis method, 134
stress/strength interference, 131–38
  application
    rationale for, 131
    target point of, 132
    timing of, 131
  benefits of, 132–34
  implementation, 132
    technique, 132
  Weibull distribution analysis technique,
    138–42
  application
    rationale for, 138
    target point of, 138
    timing of, 138
  benefits of, 142
  implementation, 139–42
    technique, 139
test data analysis, 114–27
  attribute type data, 127–31

application
  rationale for, 127
  target point of, 127–28
  timing of, 127
basic proportion technique, 128
binomial distribution technique, 128–30
implementation, 128
  technique, 128
variable type data, 114–27
application
  rationale for, 114
  target point of, 114
  timing of, 114
benefits of, 127
implementation, 114–27
  technique, 114
Variation simulation analysis, development, 80
  application
    target point for, 80
    timing of, 80
  benefits of, 83–84
  implementation
    individual responsible for, 80
    technique, 81–86
  selection of, 80
Verification test sequence, example, 47

W

Weibull distribution analysis, 138–42
  application
    rationale for, 138
    target point of, 138
    timing of, 138
  benefits of, 142
  implementation, 139–42
    technique, 139
Womack, J.P., 13
Work standards, elimination of, quality
      improvement with, 19

READER FEEDBACK
Fax to ASQ Quality Press Acquisitions: 414–272–1734

Comments and Areas for Improvement:
Practical Product Assurance Management

Please give us your comments, feedback, and suggestions for making this book more useful. We believe in the importance of continuous improvement and in meeting your needs. Your comments will help determine what improvements can be made in all ASQ Quality Press books.

Please share your opinion by circling the number below:

Ratings of the Book	Needs Work		Satisfactory		Excellent	Comments
Structure, flow, and logic	1	2	3	4	5	
Content, ideas, and information	1	2	3	4	5	
Style, clarity, ease of reading	1	2	3	4	5	
Held my interest	1	2	3	4	5	
Met my overall expectations	1	2	3	4	5	

I read the book because:

The best part of the book was:

The least satisfactory part of the book was:

Other suggestions for improvement:

General comments:

Name/Address/Phone: (optional)

Thank you for your feedback. If you do not have access to a fax machine, please mail this form to:
ASQ Quality Press, 611 East Wisconsin Avenue, P.O. Box 3005, Milwaukee, WI 53201-3005   Phone: 414-272-8575